WITHDRAWN

PROGRESS IN

Nucleic Acid Research and Molecular Biology

Volume 25

PROGRESS IN
Nucleic Acid Research and Molecular Biology

edited by

WALDO E. COHN

Biology Division
Oak Ridge National Laboratory
Oak Ridge, Tennessee

Volume 25

1981

ACADEMIC PRESS

A Subsidiary of Harcourt Brace Jovanovich, Publishers

New York   London   Toronto   Sydney   San Francisco

COPYRIGHT © 1981, BY ACADEMIC PRESS, INC.
ALL RIGHTS RESERVED.
NO PART OF THIS PUBLICATION MAY BE REPRODUCED OR
TRANSMITTED IN ANY FORM OR BY ANY MEANS, ELECTRONIC
OR MECHANICAL, INCLUDING PHOTOCOPY, RECORDING, OR ANY
INFORMATION STORAGE AND RETRIEVAL SYSTEM, WITHOUT
PERMISSION IN WRITING FROM THE PUBLISHER.

ACADEMIC PRESS, INC.
111 Fifth Avenue, New York, New York 10003

United Kingdom Edition published by
ACADEMIC PRESS, INC. (LONDON) LTD.
24/28 Oval Road, London NW1 7DX

LIBRARY OF CONGRESS CATALOG CARD NUMBER: 63–15847

ISBN 0–12–540025–X

PRINTED IN THE UNITED STATES OF AMERICA

81 82 83 84   9 8 7 6 5 4 3 2 1

Contents

LIST OF CONTRIBUTORS vii

ABBREVIATIONS AND SYMBOLS ix

SOME ARTICLES PLANNED FOR FUTURE VOLUMES xiii

Splicing of Viral mRNAs

Yosef Aloni

I.	Introduction	1
II.	SV40 as a Model System	2
III.	The Initiation of Transcription of SV40 DNA Late after Infection	4
IV.	The SV40 Minichromosome	5
V.	Splicing of "Late" mRNAs of SV40	7
VI.	Mapping the Leader and the Body of the Viral mRNAs by Electron Microscopy	9
VII.	Models for the Joining of the Leader to the Coding Sequences	14
VIII.	Techniques for Analyzing Spliced RNAs	15
IX.	Splicing of Late mRNAs of Polyoma	16
X.	Splicing of the RNA of the Minute Virus of Mice	17
XI.	Splicing of Moloney Murine Leukemia Virus RNA	21
XII.	Models for Splicing of mRNA	25
XIII.	Splicing Intermediates	27
XIV.	Conclusion	28
	References	28

DNA Methylation and Its Possible Biological Roles

Aharon Razin and Joseph Friedman

I.	Introduction	33
II.	Methylases and Their Specificity	34
III.	Distribution of Methylated Bases along the Chromosome	38
IV.	The Mode of Methylation *in Vivo*	41
V.	Possible Functions of Methylated Bases in DNA	43
VI.	Conclusions and Prospects	49
	References	50

Mechanisms of DNA Replication and Mutagenesis in Ultraviolet-Irradiated Bacteria and Mammalian Cells

Jennifer D. Hall and David W. Mount

I.	Introduction	54
II.	DNA Synthesis in Ultraviolet-Irradiated Bacteria	60
III.	DNA Synthesis in Ultraviolet-Irradiated Mammalian Cells	75
IV.	Mutagenesis by Ultraviolet Radiation	100
V.	Mechanism for Reactivation of Ultraviolet-Damaged Viruses	108
VI.	Effects of Ultraviolet Irradiation on DNA Synthesis *in Vitro*	116
VII.	Summary and Future Perspectives	121
	References	122
	Note Added in Proof	126

The Regulation of Initiation of Mammalian Protein Synthesis

Rosemary Jagus, W. French Anderson, and Brian Safer

	Introduction	128
I.	Importance of Initiation in the Regulation of Protein Synthesis in Mammalian Tissues	129
II.	Sequence of Events	133
III.	Regulation of Initiation	153
IV.	Summary: Overview on Present Understanding of the Control of Initiation	175
	References	177

Structure, Replication, and Transcription of the SV40 Genome

Gokul C. Das and Salil K. Niyogi

I.	Introduction	187
II.	Structure of the SV40 Genome	189
III.	Replication of the SV40 Genome	199
IV.	Transcription of the SV40 Genome	211
V.	The Minichromosome—A Model for the Structure and Function of Eukaryotic Chromatin	228
	References	232
	Note Added in Proof	240

INDEX 243

CONTENTS OF PREVIOUS VOLUMES 249

List of Contributors

Numbers in parentheses indicate the pages on which the authors' contributions begin.

YOSEF ALONI (1), *Department of Genetics, Weizmann Institute of Science, Rehovot, Israel*

W. FRENCH ANDERSON (127), *Laboratory of Molecular Hematology, National Heart, Lung, and Blood Institute, National Institutes of Health, Bethesda, Maryland 20205*

GOKUL C. DAS* (187), *The University of Tennessee –Oak Ridge Graduate School of Biomedical Sciences and Biology Division, Oak Ridge National Laboratory, Oak Ridge, Tennessee 37830*

JOSEPH FRIEDMAN† (33), *Department of Cellular Biochemistry, The Hebrew University Hadassah Medical School, Jerusalem, Israel*

JENNIFER D. HALL (53), *Department of Cellular and Developmental Biology, College of Liberal Arts, University of Arizona, Tucson, Arizona 85721*

ROSEMARY JAGUS (127), *Laboratory of Molecular Hematology, National Heart, Lung, and Blood Institute, National Institutes of Health, Bethesda, Maryland 20205*

DAVID W. MOUNT (53), *Department of Microbiology, College of Medicine, University of Arizona, Tucson, Arizona 85721*

SALIL K. NIYOGI (187), *The University of Tennessee –Oak Ridge Graduate School of Biomedical Sciences and Biology Division, Oak Ridge National Laboratory, Oak Ridge, Tennessee 37830*

AHARON RAZIN (33), *Department of Cellular Biochemistry, The Hebrew University Hadassah Medical School, Jerusalem, Israel*

BRIAN SAFER (127), *Laboratory of Molecular Hematology, National Heart, Lung, and Blood Institute, National Institutes of Health, Bethesda, Maryland 20205*

---

* Present address: Laboratory of Biology of Viruses, NIAID, National Institutes of Health, Bethesda, Maryland 20014.

† Present address: Biology Division, Oak Ridge National Laboratory, Box Y, Oak Ridge, Tennessee 37830.

Abbreviations and Symbols

All contributors to this Series are asked to use the terminology (abbreviations and symbols) recommended by the IUPAC-IUB Commission on Biochemical Nomenclature (CBN) and approved by IUPAC and IUB, and the Editor endeavors to assure conformity. These Recommendations have been published in many journals (1, 2) and compendia (3) in four languages and are available in reprint form from the Office of Biochemical Nomenclature (OBN), as stated in each publication, and are therefore considered to be generally known. Those used in nucleic acid work, originally set out in section 5 of the first Recommendations (1) and subsequently revised and expanded (2, 3), are given in condensed form (I-V) below for the convenience of the reader. Authors may use them without definition, when necessary.

I. Bases, Nucleosides, Mononucleotides

1. *Bases* (in tables, figures, equations, or chromatograms) are symbolized by Ade, Gua, Hyp, Xan, Cyt, Thy, Oro, Ura; Pur = any purine, Pyr = any pyrimidine, Base = any base. The prefixes S-, H_2, F-, Br, Me, etc., may be used for modifications of these.

2. *Ribonucleosides* (in tables, figures, equations, or chromatograms) are symbolized, in the same order, by Ado, Guo, Ino, Xao, Cyd, Thd, Ord, Urd (Ψrd), Puo, Pyd, Nuc. Modifications may be expressed as indicated in (1) above. Sugar residues may be specified by the prefixes r (optional), d (=deoxyribo), a, x, l, etc., to these, or by two three-letter symbols, as in Ara-Cyt (for aCyd) or dRib-Ade (for dAdo).

3. *Mono-, di-, and triphosphates of nucleosides* (5') are designated by NMP, NDP, NTP. The N (for "nucleoside") may be replaced by any one of the nucleoside symbols given in II-1 below. 2'-, 3'-, and 5'- are used as prefixes when necessary. The prefix d signifies "deoxy." [Alternatively, nucleotides may be expressed by attaching P to the symbols in (2) above. Thus: P-Ado = AMP; Ado-P = 3'-AMP] cNMP = cyclic 3':5'-NMP; Bt_2cAMP = dibutyryl cAMP, etc.

II. Oligonucleotides and Polynucleotides

1. Ribonucleoside Residues

(a) Common: A, G, I, X, C, T, O, U, Ψ, R, Y, N (in the order of I-2 above).

(b) Base-modified: sI or M for thioinosine = 6-mercaptopurine ribonucleoside; sU or S for thiouridine; brU or B for 5-bromouridine; hU or D for 5,6-dihydrouridine; i for isopentenyl; f for formyl. Other modifications are similarly indicated by appropriate *lower-case* prefixes (in contrast to I-1 above) (2, 3).

(c) Sugar-modified: prefixes are d, a, x, or l as in I-2 above; alternatively, by *italics* or boldface type (with definition) unless the entire chain is specified by an appropriate prefix. The 2'-O-methyl group is indicated by *suffix* m (e.g., -Am- for 2'-O-methyladenosine, but -mA- for 6-methyladenosine).

(d) Locants and multipliers, when necessary, are indicated by superscripts and subscripts, respectively, e.g., -m_2^6A- = 6-dimethyladenosine; -s^4U- or -^{4}S- = 4-thiouridine; -ac^4Cm- = 2'-O-methyl-4-acetylcytidine.

(e) When space is limited, as in two-dimensional arrays or in aligning homologous sequences, the prefixes may be placed *over the capital letter*, the suffixes *over the phosphodiester symbol*.

2. Phosphoric Residues [left side = 5', right side = 3' (or 2')]

(a) Terminal: p; e.g., pppN... is a polynucleotide with a 5'-triphosphate at one end; Ap is adenosine 3'-phosphate; C > p is cytidine 2':3'-cyclic phosphate (1, 2, 3); p < A is adenosine 3':5'-cyclic phosphate.

(b) Internal: hyphen (for known sequence), comma (for unknown sequence); unknown sequences are enclosed in parentheses. E.g., pA-G-A-C(C_2,A,U)A-U-G-C > p is a sequence with a (5′) phosphate at one end, a 2′:3′-cyclic phosphate at the other, and a tetranucleotide of unknown sequence in the middle. (**Only codon triplets should be written without some punctuation separating the residues.**)

3. Polarity, or Direction of Chain

The symbol for the phosphodiester group (whether hyphen or comma or parentheses, as in 2b) represents a 3′-5′ link (i.e., a 5′ ... 3′ chain) unless otherwise indicated by appropriate numbers. "Reverse polarity" (a chain proceeding from a 3′ terminus at left to a 5′ terminus at right) may be shown by numerals or by right-to-left arrows. Polarity in any direction, as in a two-dimensional array, may be shown by appropriate rotation of the (capital) letters so that 5′ is at left, 3′ at right when the letter is viewed right-side-up.

4. Synthetic Polymers

The complete name or the appropriate group of symbols (see II-1 above) of the repeating unit, **enclosed in parentheses if complex or a symbol,** is either (a) preceded by "poly," or (b) followed by a subscript "n" or appropriate number. **No space follows "poly"** (2, 5).

The conventions of II-2b are used to specify known or unknown (random) sequence, e.g., polyadenylate = poly(A) or A_n, a simple homopolymer;

poly(3 adenylate, 2 cytidylate) = poly(A_3C_2) or $(A_3,C_2)_n$, an *irregular* copolymer of A and C in 3:2 proportions;

poly(deoxyadenylate-deoxythymidylate) = poly[d(A-T)] or poly(dA-dT) or $(dA-dT)_n$ or $d(A-T)_n$, an *alternating* copolymer of dA and dT;

poly(adenylate,guanylate,cytidylate,uridylate) = poly(A,G,C,U) or $(A,G,C,U)_n$, a random assortment of A, G, C, and U residues, proportions unspecified.

The prefix copoly or oligo may replace poly, if desired. The subscript "n" may be replaced by numerals indicating actual size, e.g., $A_n \cdot dT_{12-18}$.

III. Association of Polynucleotide Chains

1. *Associated* (e.g., H-bonded) chains, or bases within chains, are indicated by a *center dot* (not a hyphen or a plus sign) separating the *complete* names or symbols, e.g.:

poly(A) · poly(U)   or $A_n \cdot U_m$
poly(A) · 2 poly(U)   or $A_n \cdot 2U_m$
poly(dA-dC) · poly(dG-dT)   or $(dA-dC)_n \cdot (dG-dT)_m$.

2. *Nonassociated* chains are separated by the plus sign, e.g.:

2[poly(A) · poly(U)] → poly(A) · 2 poly(U) + poly(A)
or $2[A_n \cdot U_m] \to A_n \cdot 2U_m + A_n$.

3. Unspecified or unknown association is expressed by a comma (again meaning "unknown") between the completely specified chains.

Note: In all cases, each chain is completely specified in one or the other of the two systems described in II-4 above.

IV. Natural Nucleic Acids

RNA	ribonucleic acid or ribonucleate
DNA	deoxyribonucleic acid or deoxyribonucleate
mRNA; rRNA; nRNA	messenger RNA; ribosomal RNA; nuclear RNA
hnRNA	heterogeneous nuclear RNA
D-RNA; cRNA	"DNA-like" RNA; complementary RNA

ABBREVIATIONS AND SYMBOLS

mtDNA	mitochondrial DNA
tRNA	transfer (or acceptor or amino-acid-accepting) RNA; replaces sRNA, which is not to be used for any purpose
aminoacyl-tRNA	"charged" tRNA (i.e., tRNA's carrying aminoacyl residues); may be abbreviated to AA-tRNA
alanine tRNA or tRNA^{Ala}, etc.	tRNA normally capable of accepting alanine, to form alanyl-tRNA, etc.
alanyl-tRNA or alanyl-tRNA^{Ala}	The same, with alanyl residue covalently attached. [*Note:* fMet = formylmethionyl; hence tRNA^{fMet}, identical with tRNA$_f^{Met}$]

Isoacceptors are indicated by appropriate subscripts, i.e., tRNA$_1^{Ala}$, tRNA$_2^{Ala}$, etc.

V. Miscellaneous Abbreviations

P_i, PP_i	inorganic orthophosphate, pyrophosphate
RNase, DNase	ribonuclease, deoxyribonuclease
t_m (not T_m)	melting temperature (°C)

Others listed in Table II of Reference 1 may also be used without definition. No others, with or without definition, are used unless, in the opinion of the editor, they increase the ease of reading.

Enzymes

In naming enzymes, the 1978 recommendations of the IUB Commission on Biochemical Nomenclature (4) are followed as far as possible. At first mention, each enzyme is described *either* by its systematic name *or* by the equation for the reaction catalyzed *or* by the recommended trivial name, followed by its EC number in parentheses. Thereafter, a trivial name may be used. Enzyme names are not to be abbreviated except when the substrate has an approved abbreviation (e.g., ATPase, but not LDH, is acceptable).

REFERENCES*

1. *JBC* **241**, 527 (1966); *Bchem* **5**, 1445 (1966); *BJ* **101**, 1 (1966); *ABB* **115**, 1 (1966), **129**, 1 (1969); and elsewhere.†
2. *EJB* **15**, 203 (1970); *JBC* **245**, 5171 (1970); *JMB* **55**, 299 (1971); and elsewhere.†
3. "Handbook of Biochemistry" (G. Fasman, ed.), 3rd ed. Chemical Rubber Co., Cleveland, Ohio, 1970, 1975, Nucleic Acids, Vols. I and II, pp. 3–59.
4. "Enzyme Nomenclature" [Recommendations (1978) of the Nomenclature Committee of the IUB]. Academic Press, New York, 1979.
5. "Nomenclature of Synthetic Polypeptides," *JBC* **247**, 323 (1972); *Biopolymers* **11**, 321 (1972); and elsewhere.†

Abbreviations of Journal Titles

Journals	Abbreviations used
Annu. Rev. Biochem.	ARB
Arch. Biochem. Biophys.	ABB
Biochem. Biophys. Res. Commun.	BBRC

*Contractions for names of journals follow.

†Reprints of all CBN Recommendations are available from the Office of Biochemical Nomenclature (W. E. Cohn, Director), Biology Division, Oak Ridge National Laboratory, Box Y, Oak Ridge, Tennessee 37830, USA.

Biochemistry	Bchem
Biochem. J.	BJ
Biochim. Biophys. Acta	BBA
Cold Spring Harbor Symp. Quant. Biol.	CSHSQB
Eur. J. Biochem.	EJB
Fed. Proc.	FP
Hoppe-Seyler's Z. physiol. Chem.	ZpChem
J. Amer. Chem. Soc.	JACS
J. Bacteriol.	J. Bact.
J. Biol. Chem.	JBC
J. Chem. Soc.	JCS
J. Mol. Biol.	JMB
Nature, New Biology	Nature NB
Nucleic Acid Research	NARes
Proc. Nat. Acad. Sci. U.S.	PNAS
Proc. Soc. Exp. Biol. Med.	PSEBM
Progr. Nucl. Acid Res. Mol. Biol.	This Series

Some Articles Planned for Future Volumes

tRNA Splicing in Lower Eukaryotes
J. ABELSON AND G. KNAPP

Ribosomal RNA: Structure and Interactions with Proteins
R. BRIMACOMBE

Metabolism and Function of Cyclic Nucleotides
W. Y. CHEUNG

Accuracy of Protein Synthesis: A Reexamination of Specificity in Codon–Anticodon Interaction
H. GROSJEAN AND R. BUCKINGHAM

Mechanism of Interferon Action
G. SEN

The Regulatory Function of the 3'-Region of mRNA and Viral RNA Translation
U. LITTAUER AND H. SOREQ

Participation of Aminoacyl-tRNA Synthetases and tRNAs in Regulatory Processes
G. NASS

Queuine
S. NISHIMURA

Viral Inhibition of Host Protein Synthesis
A. SHATKIN

Ribosomal Proteins: Structure and Function
A. R. SUBRAMANIAN

RNA-Helix Destabilizing Proteins
W. SZER AND J. O. THOMAS

Splicing of Viral mRNAs

YOSEF ALONI

Department of Genetics,
Weizmann Institute of Science,
Rehovot, Israel

I.	Introduction	1
II.	SV40 as a Model System	2
III.	The Initiation of Transcription of SV40 DNA Late after Infection	4
IV.	The SV40 Minichromosome	5
V.	Splicing of "Late" mRNAs of SV40	7
VI.	Mapping the Leader and the Body of Viral mRNAs by Electron Microscopy	9
	A. Analysis of DNA · RNA Hybrids	10
	B. Analysis of R-Loop Structures	11
VII.	Models for the Joining of the Leader to the Coding Sequences	14
VIII.	Techniques for Analyzing Spliced RNAs	15
IX.	Splicing of Late mRNAs of Polyoma	16
X.	Splicing of the RNA of the Minute Virus of Mice	17
XI.	Splicing of Moloney Murine Leukemia Virus RNA	21
	A. Duplexes between M-MuLV Genomic RNA and cDNA	22
	B. Duplexes between cDNA and Poly(A)-Containing RNA of Cells Infected with M-MuLV	22
XII.	Models for Splicing of mRNA	25
XIII.	Splicing Intermediates	27
XIV.	Conclusion	28
	References	28

I. Introduction

In the prokaryotic cell, translation of the mRNA by ribosomes starts before transcription of the message from the DNA has been completed. In a eukaryotic cell, by contrast, the process of transcription occurs in the nucleus while translation takes place mainly, if not entirely, in the cytoplasm. The two operations are separated by the nuclear membrane, providing an obvious opportunity for additional processing and maturation of the primary RNA product. The various aspects of RNA processing are covered in comprehensive reviews (1–12).

In the last three years it has become apparent that understanding the process of RNA transcript maturation is one of the great frontiers in the field of eukaryotic cell physiology. The basis for this belief has been the discovery of a novel maturation mechanism termed "RNA splicing," identified first in the DNA and RNA tumor viruses (13–28) and then extended to many eukaryotic organisms (29–32). The principle of RNA splicing is that, during processing of a primary RNA transcript, internal sequences in the molecule are excised, and the remaining pieces are ligated to give rise to a continuous, translatable RNA molecule. This mechanism implies that the gene itself is split, in the sense that it contains sequences that are ultimately translated, and sequences that are removed and not translated. Gilbert (33) has referred to the intervening sequences, those sequences on the DNA that do not end up in the translatable mRNA, as "introns"; the expressed sequences, as "exons."

The purpose of this review is mainly to describe studies on the identification of RNA splicing in the biological systems of DNA and RNA viruses, and the first attempts to determine the mechanism of this process. Reviews on split genes and RNA splicing have been published elsewhere (12, 33–35).

II. SV40 as a Model System[1]

The use of animal viruses as model systems for probing the complexities of molecular control mechanisms has been particularly fruitful. It is generally felt that an understanding of genetic regulation in viruses will provide insight into similar regulatory processes in eukaryotic cells. The molecular biology of SV40 has been under intensive investigation for a number of years, and these studies have provided considerable information regarding the regulation of gene expression, in particular, transcriptional and posttranscriptional processing of mRNA (36–39).

SV40 provides several unique advantages as a model system for such studies. They include the following.

1. The viral genome is a small circular molecule (molecular weight 3.4×10^6) that contains genetic information for only five or six proteins. The DNA can be obtained in large quantities, which is imperative for many experiments in molecular biology.
2. The same RNA polymerase (polymerase II) transcribes both viral and cellular RNA.
3. The viral and cellular RNAs undergo similar posttranscriptional modifications (e.g., polyadenylylation at the 3' terminus, capping at the 5' terminus, and internal methylation).
4. A number of mutants and hybrid viruses are available for study.

[1] See also Das and Niyogi in this volume. [Ed.]

5. Transcriptional complexes are easy to obtain.
6. The entire nucleotide sequence of this virus has been determined.

The SV40 genome is comprised of early and late genes that are localized in symmetrical halves of the viral DNA (40). The segment between 0.67 and 0.17 on the map is transcribed in a counterclockwise direction prior to the onset of viral DNA replication and codes for the "early" viral proteins. The second segment (from 0.67 to 0.17) is transcribed in abundance after initiation of viral DNA replication in a clockwise direction. It encodes the information for the late proteins: VP_1, VP_2, and VP_3. The capsid proteins have been mapped approximately between 0.95 and 0.16, 0.76 and 0.97, and

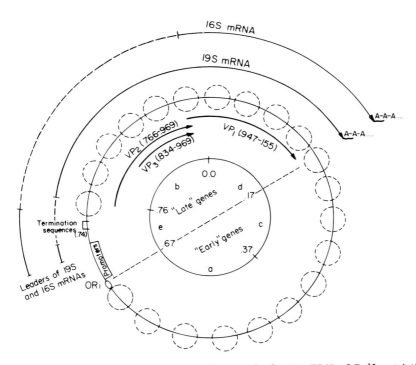

FIG. 1. Transcriptional map of SV40 indicating the five EcoRI HpaI BglI restriction fragments of SV40 genome in the central circle. Arrows on RNAs indicate 3' termini, poly(A) tails, and direction of transcription. Decimal numbers represent map units of the SV40 genome. Dashed lines (---) indicate sequences spliced out of 16 S and 19 S mRNAs. The location of the coding regions are indicated by heavy lines. The small dashed circles denote the distribution of nucleosomes about the SV40 minichromosome with an "exposed" region that contains the origin of replication (ORi) promoters for late transcription and transcription termination sequences (38). Only the major leaders of 19 S and 16 S mRNAs are represented; for more details, see ref. 38.

0.83 and 0.97, respectively (see Fig. 1) (*36 –38*). The late viral mRNAs are known as "16 S" and "19 S." The 16 S RNA codes for VP_1 while the 19 S RNAs code for VP_2 and VP_3 (*38*). The viral DNA segment between 0.67 and 0.76 is transcribed late in infection, but the genetic information encoded in this region is unknown. Our studies have been concerned with the expression of the "late" genes. Reviews concerning the "early" genes have been published (*36 –39*).

III. The Initiation of Transcription of SV40 DNA Late after Infection

The excitement that arose after the observation of RNA splicing delayed the investigations of other mechanisms involved in the regulation of SV40 gene expression. Among them are (*a*) the controls that operate at the initial steps of transcription and determine the specificity of transcriptional initiation; (*b*) the frequency of completion of primary transcripts; and (*c*) the mechanism of strand selection. I shall describe first our approaches to determine the localization of the initiation sites of SV40 late transcription, and the possible mechanisms involved in determining this specificity.

The best approach to determine the initiation site for transcription is by determining which nucleotides are at the 5′ end of the newly synthesized RNA. However, we and others (*41*) have failed to detect any labeled 5′ termini of SV40-specific RNAs. We therefore used three independent approaches, which were undertaken in order to localize the initiation site for transcription of SV40 DNA at late time after infection (*42*). Two of these were based on the Dintzis principles (*43*), while in the third we measured nascent RNA chains attached to transcriptional complexes under the electron microscope.

The rationale of localization of the initiation site(s) based on the Dintzis principles is that, after short pulses with radioactive precursors, RNA molecules would contain some labeled sequences complementary to each region of the DNA, but the labeled RNA complementary to a fragment of DNA that includes the initiation site for transcription would be in the shortest chains, while labeled RNA complementary to a DNA fragment far from the initiation site would be in successively longer chains.

RNA labeled *in vivo* was isolated from productively infected cells, and RNA labeled *in vitro* was isolated from transcriptional complexes of SV40. The purified RNAs were denatured and fractionated by sedimentation through sucrose gradients. Labeled RNAs of various lengths were hybridized with restriction fragments of SV40 DNA of a known order. In both cases, the shortest RNAs hybridized with a fragment that spans between 0.67 and 0.76 on the map. The hybridization with this fragment decreased with successively longer RNAs, indicating that transcription initiates within this fragment or very close to it. Similar enrichment for this fragment was obtained using nascent RNA chains labeled *in vitro* with a short pulse. Electron microscope observation of transcriptional complexes of SV40 has revealed a substantial fraction with one short nascent RNA chain. The initiation site of the nascent chains was placed at coordinate 0.67 ± 0.02.

Based on these and other results in which we mapped the 5' end of the nuclear viral RNA (24), we have concluded that late transcription initiates at alternative sites in a fragment of the genome that spans 0.67 to 0.76. This conclusion is supported by the localization on the map of the "caps" of the viral RNAs (44, 45) and by the localization of the 5' ends of poly(A)-containing viral RNA, using the primer extension technique (38).

The next question was: What determines the specificity of initiation? Is it only sequence-specific, or does the structure of the template contribute also to this specificity?

IV. The SV40 Minichromosome

SV40 DNA is found within infected cells in the form of a minichromosome (46), and a variety of methods have been developed for the extraction of viral chromatin from the infected cells (47–49). The SV40 minichromosome possesses a beaded structure composed of cellular histones and supercoiled viral DNA in a molecular complex very similar to that of cellular chromatin (50–52). The similarity between these structures, together with the fact that the major viral functions take place within cell nuclei via cellular machinery, have made SV40 an attractive system in which to study the organization and expression of the more complex eukaryotic chromatin. Early studies on SV40 chromatin, attempting to determine the precise distribution of the nucleosomes along the DNA, indicated that the nucleosomes were randomly distributed relative to the viral DNA sequences (53–56). Recent studies have altered this picture somewhat by indicating that a region close to the viral origin of replication is particularly sensitive to nuclease digestion (57–60). The simplest interpretation of the presence of a nuclease-sensitive region is that it contains a peculiar arrangement of protein structures about this region.

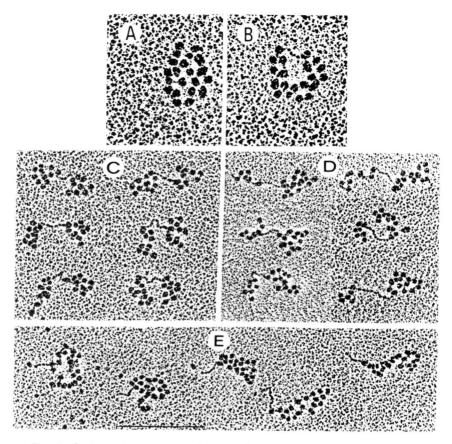

FIG. 2. Electron microscope visualization of SV40 minichromosomes. (A) Minichromosome without a gap. (B) Minichromosome with a gap. (C, D, and E) Minichromosomes with gaps cleaved with BamHI (0.15 map unit), EcoRI (0.0 map unit), and BglI (0.67 map unit), respectively.

We have analyzed SV40 minichromosomes in the electron microscope in an attempt to determine whether alteration in the gross nucleoprotein structure of this region could be visualized. About 25% of the SV40 minichromosomes observed contain a region of DNA between 0.67 and 0.75 on the map that is not organized into the typical nucleosome beaded structure (see Fig. 2) (61). This is the same region where the initiation of late transcription occurs (42) (see Fig. 1).

It was therefore of interest to investigate whether there is a correlation between the structure of the minichromosome and transcriptional initiation. As a first approach, we transcribed the viral minichromosome *in vitro* using *Escherichia coli* RNA polymerase. The initiation specificity of the bacterial RNA polymerase was highly altered by the template and initiation of transcription in the region of the major late promoters of SV40 occurs very efficiently on the viral minichromosome, but to only a minor extent, if at all, on the naked viral DNA. Furthermore, the RNA synthesized on the minichromosome elongates along the "late" strand in a clockwise direction in opposition to the "early" strand transcripts obtained from the naked DNA (62). These observations suggest that the structure of the DNA-protein template may be involved in determining transcriptional specificity.

An additional regulatory mechanism that may operate during the initial steps of SV40 late transcription has been called "attenuation" (42). According to this mechanism, there is controlled premature termination a few hundred nucleotides beyond the initiation sites. The premature termination is enhanced when the infected cells were treated with 5,6-dichloro-1-β-D-ribofuranosylbenzimidazole (63). It is interesting to note that in this region of the genome there are sequences that signal transcriptional termination (38). The site of attenuation is approximately at the 3' end of the nucleosome-depleted region (61), and is also at the site of the 3' end of the leaders of late SV40 mRNAs (see Section VI).

V. Splicing of "Late" mRNAs of SV40

The transcripts of the late strand of SV40 DNA complementary to the region between 0.70 and 0.76 on the map are abundantly represented in the cytoplasmic RNA fraction of SV40-infected cells and are also retained on oligo(dT)-cellulose (64). Since the size of the RNA containing these sequences was not determined, it was not clear that they contained a poly(A) stretch adjacent to the RNA sequences at 0.76.

To characterize the RNA transcribed from 0.67 to 0.76 of the genome, we hybridized labeled poly(A)-containing RNA from infected cells to this viral DNA fragment, eluted the RNA, and subjected it to two tests. In the first test, the ^{32}P-labeled RNA was reannealed with the five viral DNA fragments shown in Fig. 1 (central circle). In the second test, the labeled RNA was sedimented through a sucrose gradient. In a similar set of experiments, labeled poly(A)-containing RNA was hybridized to and eluted from a fragment between

0.0 and 0.17 on the map, a region of the genome known to encode the 3′ portions of the 16 S and 19 S cytoplasmic RNAs (see Fig. 1). After elution, this RNA was also analyzed both by reannealing to blots and by sucrose sedimentation. Both of the eluted [^{32}P]RNAs rehybridized in a similar pattern to the DNA fragments shown in Fig. 1. Hybridization occurred with the b (0.76–0.0), d (0.0–0.17), and e (0.67–0.76) fragments. Analysis of the eluted [^{32}P]RNAs on sucrose gradients showed that both of them contained the 16 S and 19 S species. Since the 16 S and 19 S RNAs were located previously between 0.95 and 0.17 and between 0.77 and 0.17, respectively, their hybridization with fragment e (0.67–0.76) was unexpected and prompted further investigations.

At this stage, we entertained two possibilities. The first was that we were dealing with a new species of poly(A)-containing RNA that is transcribed from 0.67 to 0.76 on the map; the second was that the RNA transcribed from this region of the genome was in some way attached to the 16 S and 19 S viral RNAs. The approach we took to distinguish between these two possibilities was to determine whether the RNA transcribed from 0.67 to 0.76 was in fact polyadenylylated and, if so, at what site. To our surprise, we found that the abundant sequences transcribed from 0.67 to 0.76 contained poly(A) at 3′ ends that were well downstream from 0.76, in fact at the same position where the poly(A)s of the 16 S and 19 S RNAs were placed on the map.

The results we collected up to this stage showed that the sequences from 0.67 to 0.76 were found in high concentrations, that they sedimented in sucrose gradients together with the 16 S and 19 S SV40 RNA species, and that they were retained on oligo(dT)-cellulose by poly(A) tails attached to 3′ ends mapping at sites not adjacent to 0.76. A model that could accommodate all these results was that the abundant sequences located between 0.67 and 0.76 were covalently linked to the 5′ ends of the coding sequences of 16 S and 19 S viral RNAs.

To test this model, we performed experiments to map the 5′ end of the 16 S RNA, taking advantage of the occurrence of a methylated cap structure at this end (44, 45). Two sets of conditions were used. In the first experiment, the methyl-labeled poly(A)-containing RNA was hybridized in formamide at 37°C to minimize thermal degradation. Under the second protocol, the labeled RNA was first fragmented to pieces 100–200 nucleotides in length and hybridized to restriction fragments at 68°C, as shown in Fig. 1. We found radioactivity associated with fragments b, d, and e in both conditions.

However, the percentage of radioactivity associated with fragment e increased about twofold and that associated with fragment b increased about 1.5-fold when the RNA was fragmented prior to hybridization. This indicated that the radioactivity associated with fragment d was in part due to nonhybridized RNA sequences covalently attached to hybridized sequences. This analysis did not discriminate between the label in cap structures and internal methyl-label.

The radiolabeled RNA associated with each of the fragments was then eluted from the nitrocellulose paper and digested with T2, T1, and pancreatic ribonucleases, and the products were analyzed on DEAE electrophoresis paper at pH 3.5. Radioactivity was found only in the cap structures and 6 mA residues (17, 20, 44, 45). Furthermore, when intact RNA was analyzed, labeled caps were found associated with fragment b (43.7%), fragment d (46.7%), and fragment e (9.6%). However, analysis of fragmented RNA showed that more than 80% of the labeled caps annealed to fragment e, and none with fragment d (17). The most plausible interpretation of these results was that the 5' end of the coding region of the 16 S poly(A)-containing RNA, already known to map from 0.95 to 0.17, covalently linked to sequences transcribed from fragment e (0.67–0.76).

We have used a different ensemble of fragments in order to locate more precisely within the EcoRI, HpaI, and BglI fragment e the sequences adjacent to the cap, and have found that the caps of 16 S and 19 S poly(A)-containing SV40 RNAs are transcribed from a region between 0.67 and 0.73 on the map. Based on these results, we have suggested a model for the splicing of 16 S viral RNA (17). According to this model, the 5'-terminal 200 ribonucleotides of late SV40 16 S mRNA are not transcribed immediately adjacent to their coding sequences. A similar conclusion was drawn for the transcription of 19 S mRNA.

VI. Mapping the Leader and the Body of the Viral mRNAs by Electron Microscopy

Confirmation for the biochemical approach, which has indicated that the leader sequences of the 16 S mRNA are not transcribed adjacent to the coding sequences, has been obtained by electron microscopy. Two types of analysis have been used. In the first, RNA · DNA hybrids were generated between the SV40 L-DNA strand, obtained by cleavage of form-I SV40 DNA with EcoRI restriction endonuclease followed by strand separation, and poly(A)-

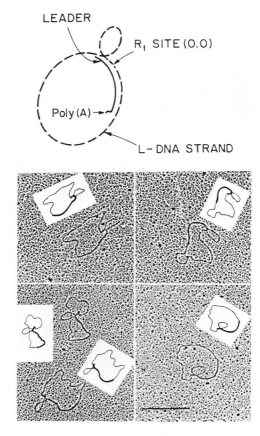

FIG. 3. Visualization of hybrids between L-DNA strands and 16 S SV40 mRNA. In the schematic illustration, the dashed line represents the L-DNA strand; the solid line, the 16 S mRNA. In the electron micrographs are some molecules scored with their tracings. The bar represents 0.5 μm.

containing RNA purified from the cytoplasm of infected cells (*18*). In the second analysis, the viral RNA was hybridized with linear double-stranded SV40 DNA (*23*). Under appropriate conditions of hybridization, individual RNA molecules can hybridize to double-stranded DNA by displacing the part of the DNA strand identical to this RNA and then hybridizing with the complementary DNA strand to form "R-loop" structures (*65*).

A. Analysis of DNA · RNA Hybrids

The typical molecule in the first analysis is circular and has two loops (Fig. 3). The small loop originates from the single-stranded

DNA that lies between the body and the leader of the 16 S mRNA. The length of this loop corresponds to the distance between the 5' end of the message and the 3' end of the leader. The large loop is partly RNA · DNA duplex corresponding to the leader and the body of the 16 S mRNA, and partly single-stranded DNA corresponding to those DNA sequences that do not code for the 16 S mRNA. Statistical analysis performed on these molecules has indicated that the small and large loops represent about 18% and 82% of unit-length SV40 DNA, respectively. The 5' end of the coding sequences of 16 S RNA appears at about 0.94 on the map; therefore, the 3' end of the leader should be 0.18 map unit from this site, at about 0.76. Measurements of the duplex regions near the small loop suggested that the length of the leader corresponds to approximately 0.04 map unit (about 200 nucleotides). This confirmed the results obtained by the biochemical methods.

B. Analysis of R-Loop Structures

R-loop structures were obtained by annealing double-stranded linear SV40 DNA (generated by cleavage of SV40 form-I DNA with *Bgl*I restriction endonuclease, which cleaves at position 0.67) with poly(A)-containing SV40 RNA, purified from the cytoplasm of infected cells late after infection. Two main groups of R-loops were observed: one represents 75% of the hybrid molecules, whose fractional length was 0.214 ± 0.010 (each value represents the mean ± SD), and the second represents 15% of the hybrid molecules, whose fractional length was 0.399 ± 0.010. The remainder of the R-loops seemed to include a minor but specific size of loop. Figure 4A shows the appearance of the most abundant R-loop structures.

We computed the location of the termini of the R-loops with reference to the *Bgl*I cleavage site at 0.67 and found that the short R-loops (Fig. 4A) span in a clockwise direction between 0.950 ± 0.007 and 0.168 ± 0.007, and the long R-loops (Fig. 5A) span in a clockwise direction between 0.770 ± 0.009 and 0.167 ± 0.010. Based on previous results, we concluded that the R-loops in Figs. 4 and 5A are of the 16 S and 19 S late mRNAs, respectively. Furthermore, the forks proximal to the short and long segments of DNA represent the map locations of the 5' and 3' ends of the bodies of these viral mRNAs, respectively. We suggested that the tails at the 3' ends represent nonhybridized poly(A) sequences, and that the tails at the 5' ends represent the leader sequences that are not coded immediately adjacent to the bodies of the messages, and are transcribed from an upstream segment in the viral DNA. We found that the frequencies of recognizable tails at the 5' and 3' ends were similar,

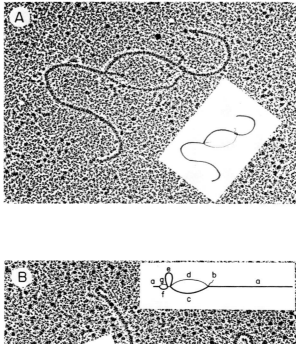

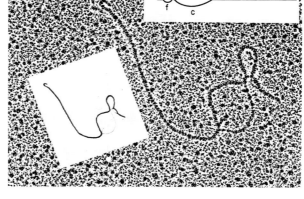

Fig. 4. Electron micrographs of R-loop structures composed of linear SV40 DNA (the superhelical form-I DNA cleaved at position 0.67 with *Bgl*I restriction endonuclease) and SV40 16 S mRNA. In (A), the leader is free; in (B), the leader formed a small R-loop. The scheme in (B) shows (right to left): a, double-stranded DNA (0.50); b, free tail of poly(A) (not measured); c, a heteroduplex DNA · RNA that includes the body of the mRNA (0.22); d, displaced single-stranded DNA (not measured); e, double-stranded DNA intervening between the body and the leader of the mRNAs (see Fig. 1) (0.18); f, a heteroduplex DNA · RNA that includes the leader of the mRNA (0.04); g, displaced single-stranded DNA (not measured); a, double-stranded DNA (0.03). The values in parentheses were obtained by measuring each segment of the molecule as the fractional length of the whole DNA. The letters in the physical map of SV40 DNA are read in a counterclockwise direction from coordinate 0.67.

SPLICING OF VIRAL mRNAS 13

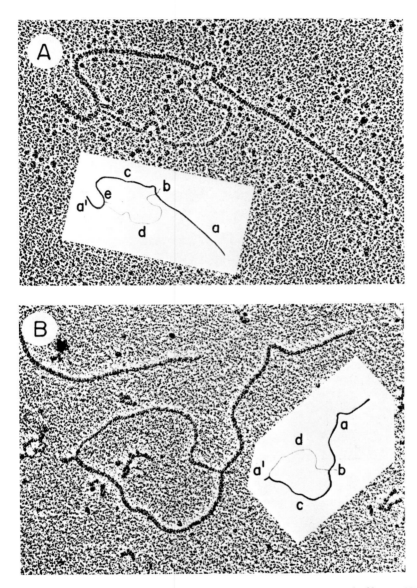

FIG. 5. Electron micrographs and their tracings of R-loops composed of linear SV40 DNA (the superhelical form-I DNA cleaved at position 0.67 map unit with *Bgl* I restriction endonuclease) and cytoplasmic 19 S mRNA in panel A or nuclear RNA-poly(A) in panel B. In the tracing, the letters represent: a, double-stranded DNA; b, free tail of poly(A); c, a heteroduplex DNA · RNA that includes the body of the RNAs; d, displaced single-stranded DNA; e, free tail of a "leader"; a', double-stranded DNA.

which indicated that all or almost all of the poly(A)-containing 16 S and 19 S late mRNAs have leader sequences at their 5′ ends. We measured the lengths of the leaders in about 100 molecules of the 16 S and 19 S mRNAs and found them to be 0.043 ± 0.012 and 0.035 ± 0.013 map units, respectively (23).

If the tails at the 5′ end represent the leader, then in a certain proportion of the molecules the sequences in the leader should hybridize to DNA sequences upstream from the locations of the bodies of the viral mRNAs and form a second R-loop with the intervening DNA segment looping out. Figure 4B shows an example of this molecule with a schematic tracing.

The R-loop analysis provided additional evidence that the leader sequences at the 5′ ends of SV40 late mRNAs are not coded at a site immediately adjacent to the main portion of the mRNAs. Moreover, the results obtained confirmed previous locations on the physical map of SV40 DNA for the main portions of the late mRNAs and established the lengths of the leaders and the map locations for their 5′ and 3′ ends (23).

VII. Models for the Joining of the Leader to the Coding Sequences

Several models for splicing of SV40 late mRNAs could be suggested. These include (a) intermolecular ligation of RNA; (b) deletion of intervening DNA sequences; (c) looping out of intervening DNA sequences so that the RNA polymerase could skip over short distances; and (d) deletion of the appropriate intervening RNA sequences.

While analyzing the transcriptional complexes of SV40, we made the following observations. First, as described above, there is one major initiation site(s) for late transcription, which maps in a fragment of the genome spanning from 0.67 to 0.76 on the map. There was no initiation site where the 5′ end of the body of 16 S RNA maps (42). This observation excluded the first possibility, which predicts two major initiation sites. The second finding was that the template for transcription is mainly the superhelical form (form I) of SV40 DNA of unit length (38, 42). This observation excluded the second alternative. We were left, therefore, with the two most attractive alternatives, in which we had to distinguish between whether splicing of the viral RNA occurs during transcription, or at the posttranscriptional level, namely on the primary RNA transcript.

For this we performed R-loop analysis on nuclear viral RNA

(24). It was shown that the nuclear poly(A)-containing viral RNA accumulates in the nucleus of the infected cell, sedimenting in sucrose gradient as a 19 S component (66).

R-loops were obtained by annealing nuclear or cytoplasmic 19 S poly(A)-containing RNA with double-stranded linear SV40 DNA obtained by cleavage of form I DNA with BglI, or TaqI restriction endonucleases (TaqI cleaves form-I DNA at 0.58). Figure 5 shows the appearance of representative R-loop structures with their schematic tracings. There are two striking differences in the appearance of the R-loops: (a) the 5' ends of the R-loops pertaining to the nuclear RNAs are closer to the restriction enzyme cleavage sites than are the 5' ends of those pertaining to the cytoplasmic RNAs; (b) free leaders are often seen at the 5' ends of the R-loops pertaining to the cytoplasmic 19 S RNA (in about 70% of the molecules), whereas they are observed much less frequently in the R-loops pertaining to the nuclear 19 S viral RNA (in about 10% of the molecules).

In an analysis of the map positions at the 5' ends of nuclear and cytoplasmic 19 S poly(A)-containing RNAs, the position of the 5' termini of the cytoplasmic RNA species was represented by a reasonably sharp histogram with a mean at coordinate 0.77 ± 0.01 on the map. The histogram for the location of the 5' termini of the nuclear viral RNAs was skewed toward an upstream position and was more heterogeneous (24).

The ability of the precursor nuclear poly(A)-containing viral RNA to anneal to the entire length of the late portion of SV40 DNA (~ 0.67 to 0.17), including the corresponding interruptions between the leaders and the bodies of the cytoplasmic mRNAs, indicates that splicing is a posttranscriptional process that occurs following polyadenylylation. Splicing of the precursor poly(A)-containing viral RNA may occur in the nucleus before the RNA molecule is transported to the cytoplasm. It is possible, however, that splicing occurs during the transport from the nucleus to the cytoplasm, or immediately after the molecule has reached the cytoplasm.

VIII. Techniques for Analyzing Spliced RNAs

In the above studies, two main techniques were described for determining the anatomy of the molecule. One is a biochemical approach, and the second is observation of RNA · DNA hybrids in the electron microscope. The biochemical approach proved to be useful because the mRNA in eukaryotic cells has tags at each end of the molecule, namely, a poly(A) at the 3' end and a methylated (cap) structure at the 5' end. The extensive information available on SV40 restriction allowed localizing the two ends of the mRNA molecule on the viral genome. This gave an estimate of the length of the RNA molecule, which was much larger than that found by direct measurement. The discrepancy between the length of the RNA molecule as determined directly and from the locations of the two ends of the molecule on the physical map of the genome can be explained only if some internal sequences in the genome between the locations of the cap and the poly(A) are not represented in the mRNA molecule. Two other widely used biochemical techniques for determining RNA splicing are the SI technique (67, 68) and the primer extension technique (38).

IX. Splicing of Late mRNAs of Polyoma

The vast amount of work carried out on investigation of the transcription of SV40 and polyoma DNAs in lytically infected cells has pointed to remarkable similarities between the two viruses. In both viruses, the stable late cytoplasmic mRNAs are usually detected as two distinct species with sedimentation coefficients of 16 S and 19 S (69–71).

In the studies on late polyoma mRNAs, we have used the same technique of R-loop mapping described for SV40 in analyzing hybrids formed between linear DNA and late poly(A)-containing cytoplasmic mRNA of polyoma. We have found that polyoma virus has three spliced late mRNAs as opposed to only two presently discovered in SV40 (25).

In the counterclockwise direction on the map, the largest R-loop spans 0.654 ± 0.013 to 0.260 ± 0.029, the middle-size R-loop spans 0.582 ± 0.013 to 0.256 ± 0.024, and the smallest R-loop spans 0.477 ± 0.020 to 0.252 ± 0.021. Based on these results and others (72), we suggested that the three types of R-loops are formed by 19 S, 18 S, and 16 S late mRNAs, respectively (25).

Figure 6 is a summary of the above results in the form of a map of polyoma DNA showing the locations of the three late polyoma RNAs and the location of the leader. Similar results have been obtained by others (73, 74).

Figure 7 shows a comparison between R-loop structures with rehybridized leaders of polyoma and SV40 late mRNAs. Double R-loops were recognizable only for SV40 mRNAs, and R-loop structures corresponding to 18 S RNA were recognizable only for polyoma late RNAs. We could not determine whether the failure to detect R-loops of 18 S mRNA in SV40 late RNAs indicates their absence in the SV40-in-

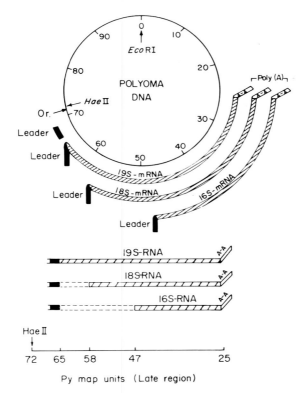

FIG. 6. Diagram of the circular polyoma DNA indicating the locations of the bodies of the 19 S, 18 S, and 16 S late RNAs and the location of the leader. The EcoRI site is taken as coordinate 0.0/1.0. One map unit corresponds to 0.01 fractional length of polyoma DNA, and the map is read clockwise. Or, is the origin of replication. The location of the 5′ end of the leader has not been determined.

fected cells, or whether they are present but at a much lower frequency than in polyoma-infected cells.

The failure to observe a second R-loop pertaining to the leader in polyoma could be explained by the observation that the leaders contain sequences that are amplified with respect to the viral genome (75).

X. Splicing of the RNA of the Minute Virus of Mice

The mechanism by which splicing is carried out has not been elucidated, and any direct approach is complicated by the elaborate splicing patterns found in the systems that have been examined so far. We

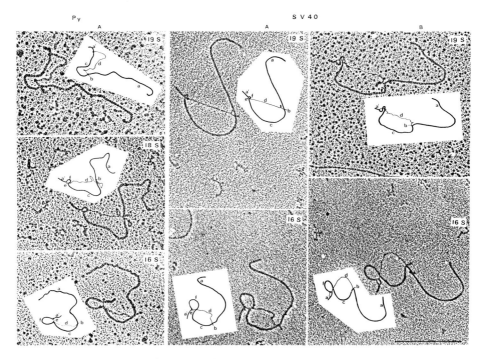

FIG. 7. A comparison between R-loop structures with rehybridized leaders of polyoma and SV40 late RNAs. For details see Fig. 4. Double R-loops were recognizable only with SV40 late mRNAs (B), and the 18 S RNA was found only in polyoma-infected cells. (Py) Scale bar represents 0.5 μm. In the tracings: a and a', double-stranded DNA; b, free tail of poly(A); c, a heteroduplex DNA · RNA that includes the body of the RNA; d, displaced single-stranded DNA; e, rehybridized leader; f, intervening DNA sequences.

have extended these studies of RNA splicing to a potentially simpler viral system, the minute virus of mice (MVM), a nondefective member of the parvovirus group. The members of this group are considered to be among the simplest of the animal viruses and comprise an icosahedral virion made from three structural protein species, containing one linear single-stranded DNA chromosome of $\sim 1.5 \times 10^6$ molecular weight (76). The MVM DNA molecule contains a stable hairpin duplex at its 5' end and a 3'-terminal structure suitable for priming complementary strand synthesis *in vitro* (77, 78). We attempted to determine the sequence arrangement of MVM RNA by examining, in the electron microscope, hybrids formed between the abundant species of nuclear and cytoplasmic poly(A)-containing RNA and single-stranded genomic DNA (28).

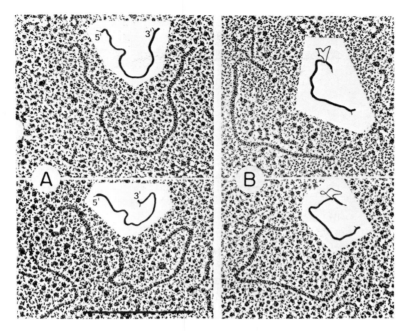

FIG. 8. Electron micrographs and their tracings of hybrids formed between nuclear (in A) or cytoplasmic (in B) RNA-poly(A) and MVM DNA. Scale bar, 0.5 μm.

Hybrids were obtained by annealing poly(A)-containing nuclear or cytoplasmic RNAs and MVM genomic DNA. Figure 8A shows the appearance of representative RNA · DNA structures with their schematic tracings. Molecules of MVM DNA of unit length were selected and examined for single- and double-stranded regions. Of more than 100 molecules of nuclear hybrid RNA · DNA scored, at least 95% exhibited a double-stranded appearance over the entire length of the DNA. It should be emphasized, however, that this analysis would not detect single-stranded regions of less than 50 nucleotides. Each hybrid examined showed a short Y-shaped structure at one end, consisting of one double-stranded and one single-stranded arm. The lengths of the two arms support the assumption that the double-stranded arm is the 130-base-pair hairpin known to be at the 5' end of the viral DNA (77), and that the single-stranded arm represents at least in part the poly(A) at the 3' end of the RNA. The opposite ends of these hybrid structures exhibit, in 50% of molecules examined, a short single-stranded region of up to 150 nucleotides, suggesting that the initiation of transcription might occur in a number of alternative sites. In this context, it is noteworthy that SV40 nuclear and cytoplasmic RNAs have various 5'-terminal sequences (see Section III above).

As the complete (or almost complete) double-stranded hybrids formed with nuclear RNA are by far the most abundant structures observed in the electron microscope, it seems that poly(A)-containing nuclear RNA is a complete (or almost complete) transcript of the genomic DNA, and suggests that it is the precursor of the cytoplasmic poly(A)-containing mRNA. Moreover, because the template for transcription is at least in part a linear DNA molecule (our unpublished results), we have concluded that transcription is initiated within a region located at the extreme 3' end of the viral chromosome.

Hybrids obtained by annealing poly(A)-containing cytoplasmic RNA and MVM genomic DNA had the appearance seen in Fig. 8B. Of about 100 molecules scored, more than 85% exhibited one short and one long arm of duplex with an intervening single-stranded DNA loop. As with the hybrids containing nuclear RNA, one end, that of the long duplex arm, has a short Y-structure comprising one double strand and one single strand, which we again postulate is the 3' end of the RNA. As with the nuclear RNA · DNA hybrids, the opposite end, that of the shorter duplex arm, exhibited a short single-stranded region in many of the molecules examined.

The observation made with the nuclear and cytoplasmic poly(A)-containing viral RNAs suggests that the MVM poly(A)-containing cytoplasmic RNA is a spliced molecule composed of two parts: a short "leader" and a longer "body." The single-stranded DNA loop (see Fig. 8B) contains the intervening template for RNA sequences, present in the nuclear poly(A)-containing RNA, and spliced out in the cytoplasmic poly(A)-containing RNA. The analysis of nuclear and cytoplasmic RNAs also indicates that, as in SV40, the sequence of modifications during mRNA maturation is polyadenylylation followed by splicing. We have measured the three parts on the RNA · DNA structures shown in Fig. 8B. Sharp histograms were obtained for each of the three parts: the leader spans $8.8 \pm 1.2\%$ of the structure; the single-stranded DNA loop (or intron), $29.3 \pm 2.4\%$; and the body, $61.8 \pm 2.3\%$.

The present study does not indicate whether the poly(A)-containing cytoplasmic RNA we have analyzed is the only viral RNA present in the infected cells. However, if there are other RNA species coded for by the viral strand, its abundance should be less than 15% of this species. Similar analysis of RNA devoid of poly(A) from infected cells revealed only a minor fraction of short heterogeneous RNA · DNA hybrids, with no accumulation of any particular length. Spliced RNA in other parvoviruses have recently been observed (79, 80).

It is thought that the MVM genome has a small number of structural genes, perhaps only one. The coding capacities of the leader and body of the poly(A)-containing RNA correspond to polypeptide molecular weights of about 12,000 and 90,000, respectively. If only the "body" codes for polypeptides, as in adenovirus and SV40 late mRNAs, it could code for the known viral protein polypeptide A (83,000 molecular weight); protein B, a second viral protein, would be processed from this. If this is true, the viral mRNA represents only about 70% of the viral genome and its primary transcript, and elucidation of the role of the remaining 30% of the primary transcript will be of great interest.

It is possible, however, that the MVM genome contains several genes, as a result of more than one functional initiation or termination site during translation. In this respect, it is noteworthy that splicing of RNA sequences at the posttranscriptional level could give rise to several functional mRNAs from the same precursor RNA, as has been found for SV40 early mRNAs (38). These mRNAs could be translated in the same reading frame, or switched from frame to frame.

The splicing of MVM RNA resembles that of the mRNAs of SV40 and polyoma, but seems to be considerably less complex. Therefore, the transcription and processing of parvovirus mRNA provides a very useful system for studying the mechanism of splicing and the physiological significance of this novel process in animal cells.

XI. Splicing of Moloney Murine Leukemia Virus RNA

RNA tumor viruses present a different viral system in which to study RNA synthesis and its processing in animal cells, since transcription can probably take place only after integration of the proviral DNA into the host chromosome. Several species of viral specific RNA have been identified in murine leukemia virus infected mouse cells. In addition to a 35 S RNA, which is similar if not identical to the genomic RNA, 20–30 S, and 14 S RNAs are the most prominent ones (*81*, *82*). The nucleic acid sequences of the smaller RNAs are probably contained within the genomic 35 S RNA, since cDNA prepared in detergent disrupted virions has been used for their detection.

In vitro translation of virus-specific RNAs revealed that the genomic 35 S RNA directs synthesis of the precursor polyprotein (Pre-76) and the reverse transcriptase, while the 22 S RNA promotes synthesis of the gp70 precursor (*82*, *83*); (gp = group protein, precursor of the envelope protein). The putative translation products of the 14 S RNA have not been characterized (*82*).

Three hypothetical models can be envisaged for the synthesis of viral RNA species smaller than 35 S: (i) each viral RNA species has its own promoter site for independent transcription; (ii) small RNA species are derived by degradation of the 5'-end portion of the 35 S RNA; (iii) a splicing mechanism is involved in the synthesis of the small viral RNAs, and thus they share a common 5'-leader sequence with the 35 S RNA. It is shown below (Section XI,B) that mechanism iii operates during production of Moloney murine leukemia virus (M-MuLV) RNA in infected cells. Three major viral RNA species with sizes corresponding to 8200, 5000, and 3000 bases have been identified with similar nucleotide sequences at both their 5' and 3' ends. The presence of a common leader sequence at their 5' ends indicates that the small 3000- and 5000-base viral RNAs are the transcript of a noncontiguous segment of DNA.

A. Duplexes between M-MuLV Genomic RNA and cDNA

Previous observations have indicated that the synthesis of cDNA, in detergent-disrupted virions, is primed by a tRNA (proline tRNA in M-MuLV) at a unique point on the 35 S genomic RNA, about 135 nucleotides from the 5' end (84–86). Furthermore, the genomic 35 S RNA contains a nucleotide sequence repeated at the 3' and 5' ends (87, 88). Such a repeated sequence probably enables the reverse transcriptase to initiate cDNA synthesis at an internal position adjacent to the 5' end of the template followed by a "jump" to the 3' end to complete copying the entire viral information. Thus, by initiating at an internal position at the 5' end, all cDNA molecules provide a "sticky" end upon reannealing with the 35 S RNA template. Hybridization of the genomic RNA with the cDNA may result in either circular or oligomeric structures. In fact, we did observe circular and long linear structures upon annealing of the genomic 35 S RNA to cDNA. Size measurement of the circular molecules indicated a unique species with circumference of 8200 bases (27). The formation of circles was dependent on hybrid formation between RNA and cDNA, because, when one of the components was omitted from the hybridization mixture, no circles were observed.

B. Duplexes between cDNA and the poly(A)-Containing RNA of Cells Infected with M-MuLV

If the small viral RNA species found in infected cells are being synthesized by a splicing mechanism, they should all contain a common leader sequence encompassing the sequences at the 5' proximal end as well as some adjacent unique sequences. Therefore, hybridization of any spliced small RNA species with a long complementary

DNA may result in a circle formed by a mechanism similar to that described for the 35 S RNA · DNA hybrid (27). Since the circumference of a circle is determined by the distance between the RNA sequences at the 5' end and at the 3' end, visualization in the electron microscope can be used as a tool to quantitate spliced RNAs in infected cells. Furthermore, a direct measurement of the circumference of the circle should provide molecular weight estimates for these RNA species.

To verify the existence of spliced viral RNA, the following experiment was performed. Poly(A)-containing RNA extracted from Swiss NIH mouse cells infected with M-MuLV was hybridized with long cDNA in excess and the RNA · DNA duplexes formed after exhaustive annealing were visualized in the electron microscope. Linear molecules (monomers, dimers, and long concatomers), in addition to circles (Fig. 9) and circles with tails (Fig. 10), were observed (27). Circularization of the smaller viral RNAs in RNA · DNA hybrids could occur only if they contain, at their 5' end, the unique sequence pres-

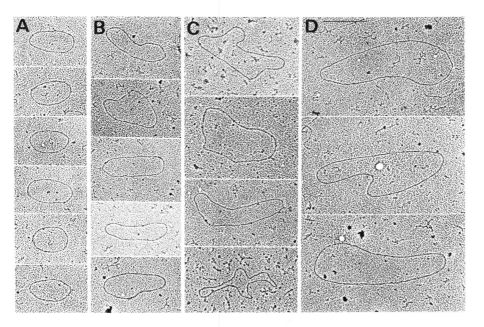

FIG. 9. Electron micrographs of circular duplexes between RNA-poly(A) of M-MuLV infected cells and cDNA. The circumferences of these circles correspond to about 3 (A), 5 (B), 8.2 (C), and 10 (D) × 10^3 bases. We suggest that the 10,000-base circles (D) and a large proportion of the 5000-base circles (B) are dimers.

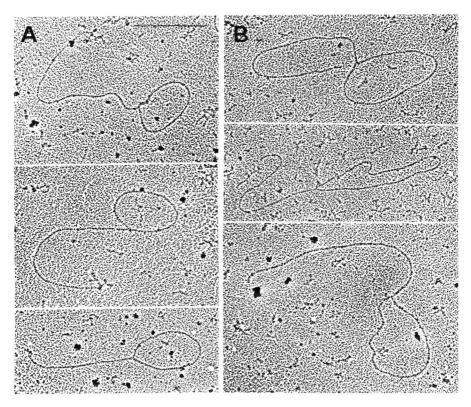

FIG. 10. Electron micrographs of hybrids between spliced M-MuLV RNAs (3000 bases in A and 5000 in B) and long cDNA. The circles are heteroduplexes of RNA · DNA, and the tails are single-stranded cDNA.

ent at the 5' end of the 35 S RNA. Circle formation is thus a direct evidence that the small viral RNAs are transcripts of a noncontiguous segment of DNA. These results provide evidence that at least two of the M-MuLV RNAs are spliced. Furthermore, based on the observation that spliced viral RNA forms a circle upon annealing with cDNA, we have suggested a new method to quantitate and determine the lengths of spliced viral RNAs in retrovirus-infected cells (27).

If the 5' end of the 35 S is spliced to the 5' ends of the smaller viral RNAs, annealing these RNAs to cDNA would form, in addition to circular structures, linear duplexes with a loop of intervening single-stranded DNA. The latter structure can be produced only if the cDNA is of unit length. Since the cDNA used in the present study was longer than 5000 bases but shorter than unit length, only the circular structures were observed. Linear duplexes with intervening single-stranded DNA loops have been observed (26).

Two types of spliced viral mRNAs have been found in cells infected with DNA tumor virus; in one type, splicing occurs within the structural gene leading to the removal of termination codons (38). In the second type, splicing occurs at the 5' end of the mRNA, and a "leader" of about 200 nucleotides is added before the AUG initiation codon (see Section VI). We could not determine from this study which of the two types of splicing occurs in M-MuLV RNA, as the lengths of the two spliced portions of the viral RNA have not yet been determined.

Biochemical techniques are currently available for the isolation and purification of circular structures from linear molecules. Thus the formation of a circular structure when retrovirus RNA is annealed with cDNA permits the isolation of the virus-specific mRNAs and their biochemical and biological characterization.

XII. Models for Splicing of mRNA

At the moment, a mechanism that allows the accurate removal of intervening sequences from a mRNA precursor and the precise ligation of the remaining sequences is unknown, and the ideas are largely speculative. Although, in the case of tRNA from yeast, an enzymic activity has been purified and characterized (89), no splicing activity has been characterized for mRNA.

Two main working hypotheses for splicing of pre-mRNA have been suggested. One is that the precursor RNA molecule itself contains the sequences and structure recognized by the enzymes involved in the splicing reaction. The second suggests that additional RNA molecules, either independently, or as a part of the splicing enzymes, provide this specificity. In either case, it is possible that the proteins attached to pre-mRNA and/or to the RNA involved in this process also contribute to the enzymic specificity.

In order to test the first possibility, several investigators have sequenced the boundaries between introns and exons (32, 38, 90, 91). High conservation of sequences spanning the splicing sites in RNA species from evolutionary distant organisms points both to a common general mechanism for processing, and to the possibility that, in a given organism, processing may be catalyzed by a single enzyme or enzyme complex.

All models for splicing predict that the ends of an intervening sequence must be brought close to each other prior to endonucleolytic cleavage, so that ligation can take place without physical separation of the 3′ and 5′ ends of the functional sequences that are to be spliced.

From sequencing these ends the following consensus emerged:

$$5' \ldots \begin{smallmatrix}A\\C\end{smallmatrix} \text{ AG } | \text{ GUAAGU} \ldots \text{UYUYYYUXCAG} | \text{G} \ldots 3'$$

$$\text{exon} \quad | \quad \text{intron} \quad\quad\quad\quad \text{intron} \quad | \quad \text{exon}$$

where Y indicates a pyrimidine nucleotide; X marks the location of a single nonconserved nucleotide in the 3′ consensus sequence, and vertical lines show the most likely location of the splice. In the position marked $\begin{smallmatrix}A\\C\end{smallmatrix}$, A occurs 11 times and C 10 times in the 26 unique sequences examined (91). The dinucleotides G-U ... A-G were found at the 5′ and 3′ ends of the introns, respectively, in almost all poly(A)-containing RNAs studied. As sequences related to this prototype sequence occur elsewhere in exons, it is unlikely to be a sufficient signal to account for splicing specificity. The intervening sequences are, however, variable in length, and possess no marked homologies or symmetries of sequence. Attempts to draw the ends of an intervening sequence together by maximizing Watson–Crick base-pairing do not generate structures in which donor and acceptor sites exhibit a reasonable constant configuration. Moreover, the signals for splicing seem to be restricted to a limited area around the junction between the intron and exon. In SV40 early mRNAs, the maximum number of nucleotides required for splicing within the coding region is 11 nucleotides on the 5′ end of the G-U donor site of small-t antigen mRNA precursor. The minimum number of nucleotides within the intron of large-T antigen mRNA precursor is not more than 12 nucleotides beyond the donor and the acceptor sites. From comparison of the nucleotide sequences of the introns in the vicinity of the large-T and small-t antigen donor sites, there is no extensive homology. Yet both of these regions are required for correct splicing (38, 92, 93). An alternative explanation is that the nucleotides flanking the splice site must be unpaired in order to provide a recognition mechanism through base-pairing with sequences remote from the splicing region, or with an independent RNA (91, 94).

Two main models have been suggested. The first predicts that the information is built into the molecule itself, and the second predicts the involvement of an additional small RNA molecule. It is argued (74) that although the primary structures of the intervening sequences are quite variable, they are rich in adenine and uridine. The adenosine and uridine residues frequently occur as homopolymeric runs of three to seven bases. On the basis of this observation and data from studies on tertiary structures of tRNA and synthetic polyribonucleotides, a model is suggested (74) in which the donor and acceptor splicing sites are brought into juxtaposition by the formation of a triple-stranded structure between the poly(A) of the mRNA precursor and regions from both the 5' and 3' ends of an intervening sequence. Support for this model comes from the observations that polyadenylylation precedes splicing (24, 28, 95), and that the histone genes that have been examined do not contain intervening sequences (96); also, histone mRNAs are not polyadenylylated (97, 98).

An independent class of small RNA that could be involved in RNA splicing is the small RNAs found in the nuclei of eukaryotic cells (99 –103). Although several of these RNAs have been sequenced, their function is unknown. They exist in cells as ribonucleoprotein particles. One complex contains seven polypeptides with molecular weights ranging between 12,000 and 35,000 (103). These complexes have been highly conserved during evolution. The 5'-terminal sequence in one major small nuclear RNA shows extensive complementarity to the consensus sequence (91). Nucleotides 3 to 8 from one small RNA known as U1 (100, 104, 105) exactly match that part of the consensus sequence found at the 5' end of the intervening sequence, while nucleotides 9 to 11 exactly match the 3' end. There are further possibilities for base-pairing. Thus it has been hypothesized that the 5'-terminal sequences of U1 RNA interact with the terminal sequences of the intervening regions and thus serve a template function by correctly aligning the sequences to be spliced. It is felt that, whatever the final outcome of these predictions, they provide an experimental framework for the study of the mechanism of RNA splicing in eukaryotic cells.

More recent studies with yeast mitochondrial RNAs point out that the normal sequence of nucleotides in an intron and its flanking sites is necessary but insufficient for the correct splice to occur. Other sequences quite remote from it are also important. The onset and progression of splicing most probably involves a temporal succession of specific three-dimensional structures that can be recognized by the enzymic machinery (106).

XIII. Splicing Intermediates

Individual intron transcripts might be removed from a precursor RNA molecule as a one-step process or in a discrete number of steps.

There is evidence that individual intron transcripts are not necessarily removed in one step in the processing of adenovirus 2 (*107*), globin (*108*), and ovalbumin nuclear RNAs (*109*). If the GU-AG rule also applies to splicing for stepwise removal of a given intron transcript, it is clear that the sequence AGGT must be present in the transcript at the intermediate splicing point. Chambon and his colleagues (*109*) have shown that such a tetranucleotide was indeed found in introns A, C, and E and was absent in introns B and D of the ovalbumin gene. However, no evidence was provided to show that introns A, C, and E are removed stepwise, while introns B and D are removed in one step.

XIV. Conclusion

It is evident that much more work will need to be carried out in order to comprehend the complex reactions involved in the proper regulation of viral and eukaryotic gene expression. The detailed studies presented here are concerned only with the overall organization of viral mRNA formation and regulation. This structuralistic analysis may provide guidelines to future studies in this exciting and fast-developing field.

Acknowledgments

The author wishes to give special thanks to Dr. S. Bratosin, among other colleagues, who made this contribution possible. This research was supported by U. S. Public Health Service Research Grant CA 14995.

References

1. J. E. Darnell, W. R. Jelinek and G. R. Molloy, *Science* **181**, 1215 (1973).
2. R. A. Weinberg, *ARB* **42**, 329 (1973).
3. E. H. Davidson and R. J. Britten, *Q. Rev. Biol.* **48**, 565 (1973).
4. G. Brawerman, *ARB* **43**, 269 (1975).
5. J. R. Greenberg, *J. Cell Biol.* **64**, 269 (1974).
6. B. Lewin, *Cell* **4**, 11 (1975).
7. B. Lewin, *Cell* **4**, 77 (1975).
8. R. P. Perry, *ARB* **45**, 605 (1976).
9. G. Molloy and L. Puckett, *Prog. Biophys. Mol. Biol.* **31**, 1 (1976).
10. L. Chan, S. E. Harris, J. M. Rosen, A. R. Means and B. W. O'Malley, *Life Sci.* **20**, 1 (1977).
11. M. Revel and Y. Groner, *ARB* **47**, 1079 (1978).
12. J. E. Darnell, This Series **22**, 327 (1979).
13. S. M. Berget, C. Moore and P. A. Sharp, *PNAS* **74**, 3171 (1977).
14. R. E. Gelinas and R. J. Roberts, *Cell* **11**, 533 (1977).
15. L. Chow, R. E. Gelinas, J. Broker and R. J. Roberts, *Cell* **11**, 819 (1977).
16. D. F. Klessig, *Cell* **12**, 9 (1977).
17. Y. Aloni, R. Dhar, O. Laub, M. Horowitz and G. Khoury, *PNAS* **74**, 3686 (1977).

18. Y. Aloni, S. Bratosin, R. Dhar, O. Laub, M. Horowitz and G. Khoury, *CSHSQB* **43**, 559 (1978).
19. M. Celma, R. Dhar, J. Pan and S. Weissman, *NARes* **4**, 2549 (1977).
20. S. Lavi and Y. Groner, *PNAS* **74**, 5323 (1977).
21. M.-T. Hsu and S. Ford, *PNAS* **74**, 4982 (1977).
22. G. Haegeman and W. Fiers, *Nature* **273**, 70 (1978).
23. S. Bratosin, M. Horowitz, O. Laub and Y. Aloni, *Cell* **13**, 785 (1978).
24. M. Horowitz, O. Laub, S. Bratosin and Y. Aloni, *Nature* **275**, 558 (1978).
25. M. Horowitz, S. Bratosin and Y. Aloni, *NARes* **5**, 4663 (1978).
26. E. Rothenberg, D. J. Donoghue and D. Baltimore, *Cell* **13**, 435 (1978).
27. A. Panet, M. Gorecki, S. Bratosin and Y. Aloni, *NARes* **5**, 3219 (1978).
28. J. Tal, D. Ron, P. Tattersall, S. Bratosin and Y. Aloni, *Nature* **279**, 649 (1979).
29. A. J. Jeffreys and R. A. Flavell, *Cell* **12**, 1097 (1977).
30. S. Tilghman, D. C. Riemier, J. G. Seidman, B. M. Peterlin, M. Sullivan, J. V. Maizel and P. Leder, *PNAS* **78**, 125 (1978).
31. C. Brack and J. Tonegawa, *PNAS* **74**, 5652 (1977).
32. S. Tonegawa, A. M. Maxam, R. Tizard, O. Bernhard and W. Gilbert, *PNAS* **75**, 1485 (1978).
33. W. Gilbert, *Nature* **271**, 501 (1978).
34. F. Crick, *Science* **204**, 264 (1979).
35. S. J. Flint, *Am. Sci.* 67, 300 (1979).
36. T. J. Kelly and D. Nathans, *Adv. Virus Res.* **21**, 86 (1977).
37. N. H. Acheson, *Cell* **8**, 1 (1976).
38. P. Lebowitz and S. M. Weissman, *Curr. Top. Microbiol. Immunol.* **87**, 43 (1979).
39. G. C. Fareed and D. Davoli, *ARB* **46**, 471 (1977).
40. G. Khoury, P. Howley, D. Nathans and M. A. Martin, *J. Virol.* **15**, 433 (1975).
41. F. J. Ferdinand, M. Brown and G. Khoury, *Virology* **78**, 150 (1977).
42. O. Laub, S. Bratosin, M. Horowitz and Y. Aloni, *Virology* **92**, 310 (1979).
43. H. M. Dintzis, *PNAS* **47**, 247 (1961).
44. D. Canaani, C. Kahana, A. Mukamel and Y. Groner, *PNAS* **76**, 3078 (1979).
45. Y. Aloni, R. Dhar and G. Khoury, *J. Virol.* **32**, 52 (1979).
46. J. D. Griffith, *Science* **187**, 1202 (1975).
47. M. White and R. Eason, *J. Virol.* **8**, 363 (1971).
48. M. H. Green, H. J. Miller and S. Hedler, *PNAS* **68**, 1032 (1971).
49. E. B. Jakobovits and Y. Aloni, *Virology* **102**, 107 (1980).
50. G. Felsenfeld, *Nature* **271**, 115 (1978).
51. R. D. Kornberg, *ARB* **46**, 931 (1977).
52. P. Chambon, *CSHSQB* **42**, 1209 (1977).
53. C. Cremisi, F. Pignatti, O. Croissant and M. Yaniv, *J. Virol.* **17**, 204 (1976).
54. C. Cremisi, P. F. Pignatti and M. Yaniv, *BBRC* **73**, 548 (1976).
55. B. A. J. Ponder and L. V. Crawford, *Cell* **11**, 35 (1977).
56. B. Polisky, and B. McCarthy, *PNAS* **72**, 2895 (1975).
57. A. J. Varshavsky, O. H. Sundin and M. J. Bohn, *NARes* **5**, 3469 (1978).
58. W. A. Scott and D. J. Wigmore, *Cell* **15**, 1511 (1978).
59. O. Sundin and A. Varshavsky, *JMB* **132**, 535 (1979).
60. W. Waldeck, B. Fohring, K. Chowdhury, P. Gruss and G. Sauer, *PNAS* **75**, 5964 (1978).
61. E. B. Jakobovits, S. Bratosin and Y. Aloni, *Nature* **285**, 263 (1980).
62. E. B. Jakobovits, S. Saragosti, M. Yaniv and Y. Aloni, *PNAS* (in press, 1980).

63. O. Laub, E. B. Jakobovits and Y. Aloni, *PNAS* **77**, 3297 (1980).
64. R. Dhar, B. S. Zain, S. M. Weissman, J. Pan and K. N. Subramanian, *PNAS* **71**, 371 (1974).
65. M. Thomas, R. L. White and R. W. Davis, *PNAS* **73**, 2069 (1976).
66. Y. Aloni, *CSHSQB* **39**, 165 (1974).
67. A. J. Berk and P. A. Sharp, *PNAS* **75**, 1274 (1978).
68. J. Favalora, R. Treisman and R. Kamen, *in* "Methods in Enzymology" (in press).
69. N. P. Salzman and G. Khoury, *Compr. Virol.* **3**, 63 (1974).
70. J. F. Sambrook, *in* "Control in Virus Multiplication" (J. Tooze, ed.), p. 153. Cambridge Univ. Press, London and New York, 1975.
71. B. E. Griffin and M. Fried, *Methods Cancer Res.* **12**, 49 (1976).
72. S. G. Sidell and A. E. Smith, *J. Virol.* **27**, 427 (1978).
73. R. Kamen, J. Favalora and J. Raker, *J. Virol.* **33**, 637 (1980).
74. H. Manor, M. Wu, N. Baran and N. Davidson, *J. Virol.* **32**, 293 (1979).
75. S. Legon, A. J. Flavell, A. Cowie and R. Kamen, *Cell* **16**, 373 (1979).
76. P. Tattersall, *in* "Replication of Mammalian Parvoviruses" (D. Ward and P. Tattersall, eds.), p. 53. Cold Spring Harbor Press, 1978.
77. G. J. Bourguignon, P. Tattersall and D. C. Ward, *J. Viol.* **20**, 290 (1976).
78. M. B. Chow and D. C. Ward, *in* "Replication of Mammalian Parvoviruses" (D. Ward and P. Tattersall, eds.), p. 205. Cold Spring Harbor Press, 1978.
79. M. R. Green, R. M. Lebovitz and R. G. Roeder, *Cell* **17**, 967 (1979).
80. C. A. Laughlin, H. Westphal and B. Carter, *PNAS* **76**, 5567 (1979).
81. H. Fan and D. Baltimore, *JMB* **80**, 93 (1973).
82. A. L. J. Gielkens, D. Van Zaane, H. P. J. Bloemers and H. Bloemendal, *PNAS* **73**, 356 (1976).
83. H. Oppermann, J. M. Bishop, H. E. Varmus and L. Levintow, *Cell* **12**, 993 (1978).
84. W. A. Haseltine, D. G. Kleid, A. Panet, E. Rothenberg and D. Baltimore, *JMB* **106**, 109 (1976).
85. J. M. Taylor and R. Ollmensee, *J. Virol.* **16**, 553 (1975).
86. G. Peters, F. Harada, J. E. Dahlberg, A. Panet, W. A. Haseltine and D. Baltimore, *J. Virol.* **21**, 1031 (1977).
87. D. E. Schwartz, P. C. Zamecnik and H. L. Weith, *PNAS* **74**, 994 (1977).
88. W. A. Haseltine, A. M. Maxam and W. Gilbert, *PNAS* **74**, 989 (1977).
89. J. Abelson, *ARB* **48**, 1035 (1979).
90. R. Breathnach, C. Benoist, K. O'Hare, F. Cannon and P. Chambon, *PNAS* **75**, 4853 (1978).
91. M. R. Lerner, J. A. Boyle, S. M. Mount, S. L. Wolwin and J. A. Steitz, *Nature* **283**, 220 (1980).
92. B. Thimmappaya and T. Shenk, *J. Virol.* **30**, 668 (1979).
93. G. Volckaert, J. Feunteun, L. V. Crawford, P. Berg and W. Fiers, *J. Virol.* **30**, 674 (1979).
94. M. Bina, R. J. Feldman and R. G. Deeley, *PNAS* **77**, 1278 (1980).
95. C.-J. Lai, R. Dhar and G. Khoury, *Cell* **14**, 491 (1978).
96. L. H. Kedes, *ARB* **48**, 837 (1979).
97. M. Adesnick and J. Darnell, *JMB* **64**, 397 (1972).
98. L. H. Kedes, *Cell* **8**, 321 (1976).
99. R. Weinberg and S. Penman, *JMB* **38**, 289 (1968).
100. G. Zieve and S. Penman, *Cell* **8**, 19 (1976).
101. W. Jelinek and L. Leinwand, *Cell* **15**, 205 (1978).
102. R. Reddy, T. S. Ro-Choi, D. Henning and H. Busch, *JBC* **249**, 6486 (1974).
103. M. R. Lerner and J. A. Steitz, *PNAS* **76**, 5495 (1979).

104. T. S. Ro-Choi and H. Bush, in "The Cell Nucleus" (H. Busch, ed.), p. 151. Academic Press, New York, 1974.
105. B.-J. Benecke and S. Penman, Cell 12, 939 (1977).
106. A. Halbreich, P. Pajot, M. Foucher, C. Grandchamp and P. Slonimski, Cell 19, 321 (1980).
107. L. Chow and T. Broker, Cell 15, 497 (1978).
108. A. Kinniburgh and J. Ross, Cell 17, 915 (1979).
109. C. Benoist, K. O'Hare, R. Breathnach and P. Chambon, NARes 8, 127 (1980).

DNA Methylation and Its Possible Biological Roles

AHARON RAZIN AND
JOSEPH FRIEDMAN[1]

*Department of Cellular
Biochemistry,
The Hebrew University Hadassah
Medical School,
Jerusalem, Israel*

I. Introduction	33
II. Methylases and Their Specificity	34
A. Substrate Specificity	34
B. Sequence Specificity	35
III. Distribution of Methylated Bases along the Chromosome	38
A. Distribution with Respect to DNA Sequences	38
B. Distribution with Respect to Chromosomal Proteins	39
C. Distribution with Respect to Chromosome Ultrastructure	40
D. Methylated and Unmethylated Domains	40
IV. The Mode of Methylation *in Vivo*	41
A. Semiconservative Methylation	41
B. *De Novo* Methylation	42
C. "Origins" of Methylation	42
V. Possible Functions of Methylated Bases in DNA	43
A. Cell Differentiation and Gene Activity	43
B. Restriction and Modification	46
C. Interplay between DNA Replication and Methylation	47
D. Mutation, Recombination, and Repair	48
VI. Conclusions and Prospects	49
References	50

I. Introduction

Bases modified by methylation have been known to occur at a low frequency in DNA for more than three decades (*1*, *2*). This modification of DNA is carried out by specific methyltransferases (DNA methylases) that transfer the chemically active methyl group from S-adenosylmethionine (AdoMet) to either carbon 5 of cytosine residues or the

---

[1] Present address: Biology Division, Oak Ridge National Laboratory, Box Y, Oak Ridge, Tennessee 37830.

exocyclic amino group attached to carbon 6 of adenine residues of the DNA chain (3).

The pattern of DNA methylation is species-specific. In the DNA of some prokaryotes, a small fraction of the cytosine residues is methylated while in others only adenine residues are methylated. A third group of prokaryotic organisms contain both 5-methylcytosine (m^5Cyt) and N^6-methyladenine (m^6Ade) in their DNA (4). Eukaryotic DNA, in general, is methylated exclusively at cytosine residues (5). The species-specific patterns of methylation reflect base specificity as well as sequence specificity of the methylases involved.

The widespread occurrence of methylated bases in DNA of various organisms, the sequence specificity of the various DNA methylases and the nonrandom distribution of the methylated bases along the chromosomes strongly suggest that modified bases in DNA are biologically significant.

Considerable efforts have been made toward the elucidation of the biological function of methylation. However, only recently has a real breakthrough been made in the methodology. The introduction of restriction enzymes and DNA cloning techniques into this field of research has already proved to be very fruitful and should provide sufficient tools for the elucidation of the function of methyl groups in DNA. In the meantime, several interesting theoretical models have been developed. These models suggest that DNA methylation may play a key role in such biological processes as DNA replication and recombination, gene expression, cell differentiation, and development (6–10).

The pattern of methyl group distribution and the mode of DNA methylation in eukaryotes seem to be quite different from what is observed in prokaryotes. In addition, some of the functions displayed by methyl groups in prokaryotic DNA may not be applicable to other organisms. The restriction-modification phenomenon, for example, has so far been observed only in bacteria (11), but this accounts for only a small fraction of the methylated bases in the host DNA, leaving the role of most of the methylated bases in DNA of prokaryotes still obscure (12). Similarly, the function of m^5Cyt in eukaryotic DNA is still essentially unknown.

II. Methylases and Their Specificity

A. Substrate Specificity

DNA methylases have been isolated and purified from a variety of eukaryotic and prokaryotic organisms. All methylases, without exception, use S-adenosylmethionine as a methyl group donor. As a rule, all methylases methylate heterologous DNA more efficiently (DNA isolated from cells other than those used for the enzyme preparation), as judged by methyl-accepting capacity in an *in vitro* assay with a cell-free methylase preparation. The basis for this is the optimal *in vivo* methylation of the potentially methylatable sites in the host DNA. On the other hand, an interesting, but still not understood, divergence in the affinity of methylases for their DNA substrates concerns the secondary structure of the DNA. While prokaryotic methylases are 10- to 20-fold more active on native double-stranded DNA (*13*), some eukaryotic methylases show enhanced activity with denatured (single-stranded) DNA (*14*). This preference may be affected by ionic strength (*15*).

On a theoretical basis, two methylase activities have been postulated (*6, 7*), namely a "*de novo*" methylase, capable of methylating unmethylated sites, and a "maintenance" methylase that can meth-

$$\begin{array}{c} \text{G A T C} \\ \text{C T A G} \end{array} \xrightarrow[\text{methylase}]{\text{"de novo"}} \begin{array}{c} \overset{m}{\text{G A T C}} \\ \text{C T A G} \end{array} \xrightarrow[\text{methylase}]{\text{"maintenance"}} \begin{array}{c} \overset{m}{\text{G A T C}} \\ \text{C T A G} \\ \phantom{\text{C T A}}_m \end{array}$$

ylate hemimethylated sites. There are now good indications from *in vivo* experiments for the existence of both *de novo* and maintenance activities (*16, 16a*). However, there is no direct proof that the two activities are performed by two different enzymes. The observed preference of a certain methylase preparation for one type of DNA substrate but not another may indicate that *de novo* and maintenance methylations perform in different proportions in various organisms. Thus, the bacterial methylase preparation might be more effective in *de novo* activity and therefore prefer double-stranded DNA. On the other hand, the eukaryotic methylase preparation might be richer in maintenance activity and thus prefer hemimethylated or single-stranded DNA as substrate. It is also possible that the *de novo* and the maintenance activities are optimal under completely different conditions, and parameters such as pH and ionic strength of the medium might determine the preferred substrate (*15*). If that is the case, the two activities probably operate *in vivo* at different phases when the conditions within the cell are appropriate.

B. Sequence Specificity

Another reason for assuming more than one methylase in a cell is the sequence specificity. Sequence specificity of bacterial DNA methylases is now a well established fact. The determination of the specific sites recognized and methylated by the various methylases has now, since the introduction of restriction enzymes, become relatively easy. The large number of different restriction enzymes that recognize and cleave various specific sites (17), and the fact that most of these enzymes are dependent on the state of methylation at the cleavage site, provide a powerful tool in the investigation of the methylated sequences.

Knowing the methylated sequences, questions concerning the metabolism of the methyl groups at those sequences can now be resolved. This information should provide clues that will aid in solving the mystery of the biological role played by the methyl groups in the DNA, their distribution along the chromosome, and the methylation pattern of single genes.

Detection of methylated sequences in DNA by the use of restriction enzymes can be achieved by at least three different methods.

1. DNA extracted from a given cell is cleaved by a pair of restriction enzyme "isoschizomers" both recognizing the same sequence, but while one enzyme cleaves only when the sequence is not methylated, the other cleaves whether or not the sequence is methylated. The restriction fragments are separated by gel electrophoresis, and the extent of methylation at the specific sequence, that is recognized by the restriction enzymes is deduced from the difference in the observed restriction pattern.

2. Heterologous DNA is methylated *in vitro* by a particular isolated methylase. The methylated DNA is then cleaved with a variety of restriction enzymes, each recognizing a different sequence and cleaving only when the sequence is not methylated. Only in the case where the restriction enzyme recognizes the same sequence that was methylated by the methylase will the DNA become resistant to cleavage; thus the methylated sequence is obtained.

3. Restriction fragments obtained by cutting with different restriction enzymes can be used as substrates for *in vitro* methylation by a particular methylase. Only the restriction fragments obtained by a restriction enzyme that shares a common sequence specificity with the methylase will be refractory to methylation, and thus the sequence specificity of the methylase can be discovered.

In prokaryotes, the modification enzymes are known to be sequence-specific, and in the class II restriction-modification systems the methylases seem to methylate the same site cleaved by the restriction enzyme counterpart (11). In *Escherichia coli* strains, most DNA methylation is not associated with restriction-modification. This methylation is carried out by at least two different methylases coded by the *dam* and *mec* genes (18). Although genetic evidence has existed for some time, physical separation of the two enzymes has only recently been achieved (13, 19-21). The *dam* product methylates adenine residues within the sequence GATC and the *mec* product methylates cytosine residues at the CC^A_TGG sequence (16). In *E. coli* DNA, these sites seem to be 100% methylated on both strands, as are those sequences in extrachromosomal DNA propagated in the cell, like plasmid or phage DNA (16). From the calculated number of GATC sites in *E. coli* DNA and the number of detected m^{6}Ade residues, it appears that the methylatable sites can account for all m^{6}Ade residues observed. However, the CC^A_TGG sites in *E. coli* DNA can accommodate only 50% of the m^{5}Cyt residues unless both cytosine moieties in this sequence are methylated. Alternatively, another sequence in *E. coli* DNA must be methylated at cytosine residues. Results suggesting that the latter alternative might be correct were recently obtained by pyrimidine isostich analysis of *E. coli* MRE600 DNA (13). If two different sequences are methylated, the question arises whether the *mec* methylase methylates both sequences or two different cytosine-DNA methylases exist in *E. coli*. The results of the experiments performed with *E. coli* MRE600 and *E. coli* SK favor the second alternative (13, 21). The authors of these papers claim that two cytosine methylases can be isolated from *E. coli*, each methylating a different sequence.

Our knowledge concerning sequence specificity of eukaryotic DNA methylases is even more limited than that described above for *E. coli* methylases. With the exception of some protozoan DNAs in which m^{6}Ade was found (22), all eukaryotic DNAs contain m^{5}Cyt as the only methylated base. However, m^{5}Cyt is not limited to one specific sequence. There are several different sequences that share in common the C-G dinucleotide in which m^{5}Cyt is found. Thus, sequences like CCGG, GCGC, and ACGT are already known to be partially methylated in many eukaryotic organisms (10). Therefore, the question arises whether one methylase characterized by a wide range of sequence specificity is responsible for the methylation of all various sequences or whether, alternatively, several highly specific methylases exist in the cell, each methylating one specific sequence. The combined use of isolated methylases and restriction enzymes should provide an answer to this very important question.

Another difference in the mode of action of eukaryotic methylases as compared to prokaryotic methylases is reflected in the extent to which a potentially methylatable sequence is methylated. In contrast to prokaryotes, the methylatable sites in eukaryotic DNA are not fully methylated. The CCGG sites in the rabbit, for example, are 50% methylated (23); in the mouse, about 70% of these sites (24, 25), in humans, 60% (26), and in calf thymus DNA, 90% of the CCGG sites are methylated (27).[2]

Several recent reports demonstrate that extrachromosomal DNA in eukaryotic cells, such as rDNA in the *Xenopus* oocyte (28) or viral DNA in virus-producing cells, are not methylated, while chromosomal rDNA in somatic cells of *Xenopus* (28) or viral DNA integrated into the host chromosome, are methylated (29, 30). These observations reflect an additional difference in the mechanisms involved in the process of DNA methylation in eukaryotes compared with prokaryotes.

III. Distribution of Methylated Bases along the Chromosome

A. Distribution with Respect to DNA Sequences

There are numerous indications that methylated bases are distributed along the chromosome in a nonrandom fashion. The origin of replication of *E. coli* DNA, for example, appears to be heavily methylated. The sequences methylated in *E. coli* DNA (16) occur in extremely high frequency in the 422 base-pair fragment, which contains the origin of replication (31, 32). Thus, the sequence GATC, methylated at the adenine residue, is repeated 22 times in this fragment. This represents a 15-fold higher frequency than expected on a purely random basis. The CC_T^AGG sequence, which is methylated at least at the internal cytosine residue, appears in a 8-fold higher frequency than expected. Since essentially all methylatable sites in *E. coli* DNA are actually methylated (16), it is possible to regard the distribution of methyl groups as reflecting the organization of methylatable sequences along the chromosome. The extraordinary high concentration of methylatable sequences at the origin of replication of *E. coli* DNA indicates therefore an unusual cluster of methyl groups. The selective mechanisms by which this unusual sequence frequency evolved, and the functional significance of such a cluster

---

[2] The extent of methylation of CCGG sites faithfully reflects the extent of methylation of the C-G dinucleotide (89). Using a modification of the nearest-neighbor analysis, which permits the direct determination of methylated C-G, it was shown that more than 50% of the dinucleotide C-G in many different mammalian DNA samples is methylated. A study of *in vitro* methylation using eukaryotic methylases demonstrates that their activity is linearly dependent on the C-G content of the DNA template (89). Taken together, these results suggest that the C-G dinucleotide is a necessary and probably sufficient signal for eukaryotic DNA methylation, both *in vivo* and *in vitro*.

of methyl groups in a strategic region like an origin of replication, are very important questions that should be studied carefully.

The entire genome of the bacteriophage ϕX174 has recently been sequenced (33). Surprisingly, this single-stranded DNA contains no GATC sequences, but the sequence CC^A_TGG appears twice in the genome. Accordingly, no m^{6}Ade has been found; however, one m^{5}Cyt residue has always been observed in the progeny virus DNA (34). The one sequence always methylated is CCTGG at position 3501; the other sequence that may be methylated in only a fraction of the viral DNA molecules is CCAGG at position 882 (16, 35).

In eukaryotic DNA, the methylated cytosine residues are also nonrandomly distributed with respect to sequences of the DNA. The abundance of m^{5}Cyt in highly repetitive sequences is severalfold higher than in intermediary repetitive or unique sequences (36 –40). Methylated and unmethylated domains have been observed in sea urchin DNA (41) (Section III,D). However, since in eukaryotic DNA only a fraction of the methylatable sequences are methylated (see Section II,B), the distribution of m^{5}Cyt along the chromosome cannot be determined merely by the distribution of the methylatable sequences, and some other factor, involving perhaps chromosome structure, takes part in the determination of the methylation pattern. The advantage of such a pattern will be revealed only when the biological functions played by m^5 Cyt in eukaryotic DNA is elucidated.

B. Distribution with Respect to Chromosomal Proteins

The methylated cytosine residues are also nonrandomly distributed with respect to the nucleosomal structure of the eukaryotic chromosome. Using micrococcal nuclease to digest chromatin, it was found that m^{5}Cyt is nonrandomly distributed with respect to chromatin proteins (42). During early digestion times, when up to a fifth of the total nuclear DNA is digested, the m^{5}Cyt remains essentially resistant to digestion. This indicates that very few m^{5}Cyt residues are located in the nuclease-sensitive "spacer" regions between nucleosome cores. Since at 20% digestion most of the DNA is found in core particles, it can be concluded that a very high percentage of the methyl groups are located preferentially in core regions. This enrichment of methyl groups is not due to the presence of a specific type of m^{5}Cyt-containing sequence in these regions. When chromatin DNA after partial digestion was self-hybridized, m^{5}Cyt was found to be distributed in satellite, intermediate repetitive, and unique DNA sequences, much the same as in total DNA (39).

These results suggest that in addition to the sequence specificity of the eukaryotic methylases, DNA structure and its interaction with chromosomal proteins guide the methylase in finding the right sequences to methylate. It is well known that nucleosomes are assembled soon after DNA replication (43), but it is not known whether nucleosome assembly precedes DNA methylation. It is therefore still unclear if and how chromosomal proteins affect the pattern of methylation in eukaryotic chromosomes.

C. Distribution with Respect to Chromosome Ultrastructure

By means of anti-m^{5}Cyt antibodies, m^{5}Cyt has been located in the heterochromatic regions of the metaphase chromosome. These regions have high concentrations of satellite DNA. In the human, m^{5}Cyt clusters in the C-banding region of chromosomes 1, 9, 16, and 15, corresponding to sites where satellites II or IV and, to a lesser extent, satellites I and III are concentrated (37). In the mouse, the methylated regions correspond to the locations of satellite DNA, i.e., the C-band region of virtually every mouse chromosome (37). It should be noted that it is possible that many methyl groups that are obscured by the ultrastructure of the chromosome will escape the reaction with anti-m^{5}Cyt. Therefore, only "exposed" m^{5}Cyt residues will be detected by this technique.

D. Methylated and Unmethylated Domains

A recent study with DNA of the sea urchin *Echinus esculentus*, using the C-G restriction enzymes *Hpa*II, *Hha*I, and *Ava*I (restriction enzymes that recognize sequences containing C-G), revealed that, at least in this DNA, two compartments exist with respect to methylation (41). One compartment is heavily methylated (m+) and comprises stretches of 15,000–50,000 bases. The other compartment carries either very little or no m^{5}Cyt (m−). Similar (m+) and (m−) compartments are found in the different stages of development of the sea urchin and in DNA of different tissues of this organism. Whether this pattern is unique to sea urchin or rather universal, remains to be elucidated.

When methylatable sequences were studied in single genes by the use of C-G enzymes and "Southern blotting" (44) as was done for the chick ovalbumin gene (45), three classes of methylatable sites were found: sites methylated in all tissues (m+); sites not methylated in any of the tissues (m−); and a variable fraction of sites methylated in some tissues but not in others (m^v). Unmethylated sites of a specific gene are usually found in tissues that express the gene (see Section V,A). Therefore the distribution of the methyl groups with respect to specific gene regions might be determined early in development (Section V,A).

IV. The Mode of Methylation *in Vivo*

All the DNA sequences that we recognize as being candidates for methylation constitute palindromes with 180° rotational symmetry. Therefore, as mentioned in Section II,A, two different modes of methylation were proposed: one semiconservative type methylating symmetrically hemimethylated sequences

$$\begin{array}{c} \overset{m}{C}CGG \\ GGCC \end{array} \longrightarrow \begin{array}{c} \overset{m}{C}CGG \\ GG\underset{m}{C}C \end{array}$$

the other methylating "*de novo*" unmethylated specific sequences

$$\begin{array}{c} CCGG \\ GGCC \end{array} \longrightarrow \begin{array}{c} \overset{m}{C}CGG \\ GGCC \end{array}$$

A. Semiconservative Methylation

Replicating DNA has been shown to undergo methylation on the nascent DNA strand at or near the replication fork. This important observation has been made in prokaryotes (46, 47) as well as in eukaryotes (48, 49). Thus, it is possible that the parental DNA strand that carries the methylation pattern can serve as "template" for a semiconservative type of methylation that will copy the methylation pattern from the parental strand onto the newly synthesized DNA strand. Supporting this type of methylating mechanism are the observations made both in prokaryotes (16) and in eukaryotes (27, 49) that the methylated sites are symmetrically situated on both strands. This type of methylating activity seems to be essential for maintaining a stable clonal pattern of methylation as observed in the DNA of all organisms. A "maintenance" enzyme that will carry out this enzymic activity was postulated on theoretical grounds (6, 7). According to these models, the methylation pattern that is laid down early in development is heritably transferred from generation to generation with high fidelity by a maintenance methylase that carries out semiconservative methylation concomitant with the process of DNA replication.

There is no report in the literature of which we are aware that describes the isolation of maintenance-type enzyme, although *in vivo* experiments clearly indicate that it exists (*16a*).[3]

B. De Novo Methylation

It is self-evident that modification methylases involved in bacterial restriction-modification systems are capable of methylating *de novo* unmethylated DNA. This enzymic reaction must be slow, since very few foreign DNA molecules that penetrate bacterial cells survive nucleolytic cleavage by the restriction enzyme. Although suggested some time ago in several proposed models (*6, 7*), *de novo* methylating activity other than that of the prokaryotic modification enzymes has not yet been described.

In recent studies, *de novo* methylation has been shown to occur in prokaryotic cells (*16*). DNA of mouse mitochondria, which contains GATC sites, was inserted into *E. coli* cells by transfection with mtDNA-pBR322 recombinant plasmid. All the GATC sites in the mtDNA, which were not methylated originally, became methylated within the *E. coli* host cell (*16*). This result proves that the *dam* gene product of *E. coli* is capable of *de novo* methylation of GATC sites.

C. "Origins" of Methylation

Does the DNA methylase bind to every methylatable site, methylate it, and dissociate when the site is methylated? Or, is there a methylation origin where the methylase binds and from this initiation site "walks" along the DNA, methylates all methylatable sites on its way, and then falls off at a termination signal? Early studies with the rat liver methylase support the latter possibility (*50*).

Since in prokaryotes all sites in chromosomal and extrachromosomal DNA are methylated, it is suggested that methylation in prokaryotes is a "trans-acting" process, meaning that the methylase can "jump" from one site to another, and methylate both. Methylation in eukaryotes may, however, be a processive event that starts and

---

[3] Experiments were designed to study the mode of DNA methylation in eukaryotic cells by DNA-mediated gene transfer. Foreign DNA methylated *in vitro* was introduced into mouse tk⁻ L cells by cotransformation with the *H. simplex* tk gene. The inserted DNA was assayed for methylation at specific sequences by restriction enzyme analysis and Southern blotting. The results reveal that sequences recognized by the cell methylase, such as CCGG, remain partially methylated after more than 80 cell generations (*16a*). On the other hand, methyl groups present in sites normally not methylated in the eukaryotic cell (CC^A_TGG or GATC) are lost, probably during replication (*90*). This suggests that the pattern of methylation is clonally inherited in a semi-conservative manner.

stops at recognized initiation and termination sites. Such a mechanism would explain why sequences that are methylated when integrated into the cell DNA are found to be methylated, whereas the same sequences are not methylated as extrachromosomal DNA (30, 51).

V. Possible Functions of Methylated Bases in DNA

Although, in general, the biological significance of methylated bases in DNA is still an enigma, a number of recent experiments strongly suggest that DNA methylation is, in one way or another, involved in major biological processes. Since the methyl groups of both m^6Ade and m^5Cyt lie in the major groove of the double helix, they are exposed to the surroundings and could serve as signals affecting protein–DNA interaction. That this is indeed the case can be deduced from the fact that interaction of restriction enzymes with DNA is dependent on the state of methylation at the recognition site. Another example where a methyl group in the major groove affects a specific binding of a protein is the obligatory requirement for the presence of a methyl group in the major groove at position 13 of the lac operator, in order for the lac repressor to bind (52).

A. Cell Differentiation and Gene Activity

Several models have been proposed to explain control of gene activity based on DNA methylation. The first model suggesting that DNA methylation plays a role in cell differentiation predicts transition of $G \cdot C$ to $A \cdot T$. It was assumed that m^5Cyt residues at specific sites in the DNA undergo deamination to thymine causing an heritable change (53). The thymine residues produced by this deamination (called DNA minor thymine, or DMT) are expected to receive their methyl groups from methionine and to be nonrandomly distributed in the DNA. In addition, DMT should be found only in cells undergoing terminal differentiation. In fact, in developing sea urchin embryos, DMT was observed (54) and shown to be nonrandomly distributed (55). However, DMT was also found in Novikoff hepatoma cells (56) and in prokaryotes (34), and it was later questioned as to whether it originates from the deamination of m^5Cyt (57). In spite of attempts to find a DNA-m^5Cyt deaminase, no such enzyme has been observed. Although this model does not explain heritable changes during development and differentiation, it might help in the understanding of heritable changes on an evolutionary scale (58).

An extension of this hypothesis, into which additional elements of developmental clocks were incorporated, was published later. It postulated base transitions that require the involvement of yet unidentified enzymes like deaminases, demethylases, and aminases (7). As discussed in detail in previous sections, the postulated methylase activities are of two kinds; one methylating *de novo* unmethylated sites, and the other methylating DNA that is methylated only on one strand in a semiconservative manner (7). The enzyme methylating unmethylated specific sequences is suggested to be a very slow-acting methylase, whereas the second enzyme is rapidly methylating half-methylated sites in order to maintain the pattern of methylation through cell division. This enzyme should be present in somatic cells at least during S phase or G_2 in order to methylate hemimethylated DNA formed during replication, thereby ensuring an inherited clonal stability of the pattern of methylation. This mode of DNA methylation was used to explain the mechanism for X inactivation (6).

A more general hypothesis based on DNA methylation attempts to explain the whole set of developmental processes leading to selective "silencing" of eukaryotic DNA (8). It proposes that processes like the uniparental inheritance of DNA, chromosome elimination, heterochromatization, and X-chromosome inactivation are all regulated by the same mechanism, restriction-modification of DNA as it is known for prokaryotic systems. Only in one case, uniparental inheritance of chloroplast DNA, has it been unequivocally shown that restriction-modification of DNA is in fact controlling this process (59).

If methylation of DNA were involved in differentiation and regulation of gene expression, tissue-specific patterns of DNA methylation would be expected. However, in studies where the overall content of m^5Cyt in DNA of various tissues was determined, no significant differences in m^5Cyt content were observed (24, 42). Although in some early experiments tissue specificity was reported, the observed differences were generally around 10% (60, 61). One exception is a report on the content of m^5Cyt in sperm DNA of several vertebrates. These DNAs were found to contain about half the amount of m^5Cyt found in DNA of somatic cells (62). However, quantitation of m^5Cyt in sequences containing C-G by "C-G restriction enzymes" revealed that sperm DNA of chicken (45), rabbit (63), and human (64) is highly methylated as compared to somatic cells, whereas sea urchin sperm DNA has the same m^5Cyt content as somatic cells (48).

Even when the extent of methylation of a specific sequence like CCGG is studied in DNA from various tissues by a direct method (27), the variations observed are very small, whereas the species specificity observed is very prominent. About 50% of the CCGG sequences in rabbit DNA is methylated (23), 60% of these sequences are methylated in essentially all human tissues (26), and 75–80% in the mouse (24, 25). Tissue specificity is observed when methylation of a specific sequence (CCGG) in a unique gene is studied. Restriction fragments of genomic DNA obtained by cleavage with MspI (a restriction enzyme that cleaves whether or not CCGG is methylated at the internal C) and HpaII (a restriction enzyme isoschizomer that cleaves only unmethylated CCGG) are transferred to nitrocellulose paper. The gene-containing fragments are then visualized by hybridization and autoradiography with a labeled probe (cloned gene DNA).

The first indication of tissue specificity came from a study with the rabbit globin gene. A CCGG sequence located in the large intron of the β-globin gene was found to be 50% methylated in most somatic tissues, but 80% methylated in brain and 100% in sperm (63). This very encouraging discovery, followed by similar observations in chicken globin (65), human globin (26, 64), and chicken ovalbumin (45) genes, suggested that DNA methylation might in fact be involved in cell differentiation and gene expression.

It is clear that the specific changes in methylation pattern that occur during differentiation take place against a high background of methylated sequences, such that the average methylation level changes less than 10%.

Experiments in which methylation of sequences containing C-G within specific genes was studied revealed a remarkable correlation between the extent of methylation of certain sites in the gene and its activity. Certain CCGG sites in the chicken β-globin gene were found to be less methylated in erythrocytes than in oviduct tissue (65). Along the same line, CCGG and GCGC sites in the ovalbumin, ovotransferrin, and ovomucoid genes were found to be less methylated in the chicken oviduct than in other tissues (45).

A very detailed study of several sequences containing C-G in the various human globin genes with respect to their extent of methylation in human tissues revealed that, in general, these sequences are less methylated in the tissue that expresses the particular gene. Some exceptions to this rule also exist where, in tissue in which the gene is usually not expressed, some sequences are also undermethylated (26). The conclusion that can be drawn from these results is that undermethylation is perhaps required, but is not sufficient, for gene expression. Several similar studies with viral genes revealed essentially the same results. The majority of the CCGG sequences present in the integrated adenovirus type 12 (Ad 12) DNA are methylated. Ad 12 DNA isolated from the virion is not methylated. The segments of integrated DNA that comprise early genes, and that are expressed in two lines of transformed hamster cells, are undermethylated in comparison to viral segments coding for late functions; and vice versa, segments that contain late viral genes are less methylated than segments of early genes in two lines of Ad-12-induced rat brain tumor cells, where the late genes are expressed (30). Cell lines transformed with Herpes saimiri, but not producing virus, contain integrated copies of viral DNA. This DNA is highly methylated, but viral sequences in virus-producing lymphoid lines are not methylated (29).

It is clear that all the experiments described above do not answer the question whether undermethylation of active genes is a cause or an effect of gene expression. A promising system to answer this question might be the recently developed technique of DNA-mediated transfection of eukaryotic cells (66).

In the meantime, experiments have been performed in naturally occurring systems where cell differentiation can be induced and where the induction is accompanied by a concomitant decrease in DNA methylation. Treatment of mouse $10T^{1}/_{2}$ cells with 5-azacytidine for a short time during DNA synthesis results in reprogramming of differentiation. Several generations after the drug has been removed, foci of muscle cells, adipocytes, and chondrocytes appear in the culture (67).

In 5-azacytidine, a nitrogen atom replaces carbon-5 of the pyrimidine ring, so that enzymic methylation at this position is impossible. The fact that 5% of the cytosine residues in the DNA is replaced by 5-azacytidine suggests that the methylation pattern of the DNA is possibly altered. In addition, inhibition of DNA methylase activity in 5-azacytidine-treated cells was observed (68). This system provides a promising tool for study of the possible function of DNA methylation in differentiation and regulation of gene activity.

In other studies, the correlation between globin gene activity and the state of methylation in the cell was investigated. The globin gene was induced in erythroleukemic cells in culture by the treatment with either ethionine (69) or nicotinamide (70). Both compounds are known potentially to alter transmethylation using S-adenosylmethionine (AdoMet) as methyl donor. Ethionine is activated by ATP, and the product, S-adenosylethionine, inhibits methylase activity by competition with AdoMet. Nicotinamide is methylated by a eukaryotic specific methylase and can deplete the intracellular pool of AdoMet. But these two compounds also exhibit other affects on metabolic reactions in which methyl group transfer is involved. One major process inhibited by these reagents is protein biosynthesis. It is, therefore, impossible to draw clear-cut conclusions from these experiments, especially in light of the fact that many other compounds are known to induce globin formation in these cells without any indication that they affect DNA methylation.

B. Restriction and Modification

The only biological phenomenon for which it has been shown unambiguously that methyl groups in DNA play a role, is that of restriction and modification in bacteria (*11*, *71*). It is interesting to note that, while in most cases the presence of a methyl group in a cleavage site protects this site from being cleaved by a restriction enzyme, there are a few instances in which the interaction of the nuclease with the DNA at this site is indifferent to or even dependent on the presence of a methyl group (*16*, *72*).

In eukaryotic cells the extent of methylation of methylatable sites seldomly reaches 100%. This fact by itself makes unlikely the possibility that restriction-modification exists in eukaryotes. There is only one example in which a restriction-modification type of mechanism was reported for a eukaryotic system, the process of zygote formation of chloroplast DNA in *Chlamydomonas* (*59*). The chloroplast DNA from mating type mt+ is highly methylated, whereas DNA from mt− is slightly methylated. After zygote formation, mt+ survives and mt− is degraded. It is still questionable how typical of eukaryotic cells this system is. One successful attempt to detect a restriction-type enzyme has also been recently reported (*73*). This endonucleolytic activity was found in hamster kidney fibroblasts and exhibited bacterial restriction enzyme type I-like activity. The enzyme was dependent in its activity on the presence of AdoMet and cleaved DNA at nonspecific sites. Type II-like restriction activity has also been reported (*51*).

C. Interplay between DNA Replication and Methylation

The extremely high frequency of methyl groups in the origin of replication of *E. coli* DNA (see Section III,A) is indicative of some interrelation between the complex process of DNA replication and DNA methylation. Early studies of methionine starvation of met^- *E. coli* cells already suggested that DNA replication in *E. coli* might be dependent on the normal methylation of the DNA subsequent to its replication (*46*, *47*). In recent preliminary experiments the kinetics of DNA synthesis and DNA methylation were studied in synchronized cultures of *E. coli* (*74*). These experiments revealed a discontinuous "staircase" type of kinetics for both DNA synthesis and DNA methylation. Short intervals of DNA synthesis, during which no methylation occurs, are followed by bursts of methylation (lasting 1–2 minutes at 30°C) in which no DNA synthesis takes place. This unexpected observation indicates that DNA replication and methylation are coupled in *E. coli*.

Several studies with the bacteriophage ϕX174 suggest a function for the m^5Cyt residue found in the viral DNA (position 3501, see Section III,A) in terminating a round of DNA replication (*75*, *76*). Abnormal replicative intermediates accumulate in an infection when DNA methylation is inhibited by nicotinamide (*77*). The same abnormal intermediates are observed when cells are infected with a virus defective in gene A that codes for a specific endonuclease, which is believed to cleave one genome-length progeny DNA during phage maturation (78).

It has been proposed that the control of eukaryotic DNA replication is achieved by methylation (9). In bacteria, replication of DNA proceeds from a single origin, and reinitiation occurs in rich media even before cell division, whereas the eukaryotic chromosome is replicated from numerous origins (79). However, reinitiation does not occur in eukaryotic cells until all segments of all chromosomes are complete. Normally, reinitiation occurs only after the cell has divided and passed through G_1. This process must, therefore, be carefully regulated. Although not all initiation sites are used in one replication cycle, after the fork passes over, all sites, used as well as unused, must be modified to prevent reinitiation. This modification should be reversible to enable resumption of initiations at the beginning of S phase in the next cycle. The model for control of these initiations by DNA methylation suggests that the replication complex binding sites are methylated after replication at G_2. Another assumption is that the replication complex does not bind to half-methylated DNA. After methylation in G_2, the DNA is available for reinitiation if the proper enzymes are available. Currently available data with regard to DNA methylation in eukaryotes are not yet sufficient to support such a model. Results concerning the cell cycle stage when DNA methylation takes place are conflicting (48, 60). Better procedures for synchronizing cells in culture and mutants in various cell cycle phases are required to test the validity of the hypothesis.

D. Mutation, Recombination, and Repair

The methylated bases in DNA could serve both as "hot spots" for mutations or, in contrast, play a role in protecting DNA from being mutated. There is evidence that both functions are probably used in bacteria. In *E. coli*, a major hot spot for mutation in the *lac* repressor gene appears to be in the sequence CC_T^AGG (80). As mentioned in Section II,B, this sequence in *E. coli* DNA is completely methylated at the internal cytosine residue by the *mec* methylase.

It has been suspected for a long time that cytosine residues in DNA are subject to deamination; the product of cytosine deamination, being uracil, can be enzymatically removed (81). But deamination of 5-methylcytosine produces thymine, which must lead to a heritable change in the DNA by transition from $G \cdot C$ to $A \cdot T$ (53). In eukaryotic DNA, m^5Cyt is preferentially present in C-G sequences. This sequence is also underrepresented, to various extents, in vertebrates. The observed deficiency of the dinucleotide C-G was suggested to result from high mutability of m⁵Cyt (82). In a recent comparative study of the extent of methylation of C-G sequences in the DNAs of a large variety of eukaryotic organisms, a reciprocal correlation was observed between the content of m⁵Cyt and the frequency of the C-G dinucleotide. This observation could support the idea that methylation of C-G renders it hypermutable (58). However, if methylated C-G is indeed a hot spot for mutation, it must be quite rare, as otherwise the methylated C-G sequences would have been quickly eliminated. The fact is that the changes in C-G content are seen clearly only on an evolutionary scale.

Another methyltransferase in *E. coli*, the *dam* protein, methylates the sequence GATC at the N^6 position of the adenine residue. Studies with mutants deficient in *dam* methylase revealed that they are defective in a number of functions associated with DNA metabolism. The mutant cells show increased mutability (83) and increased levels of recombination (84). Although *dam*⁻ mutants are viable, they give rise to lethal phenotypes in conjunction with nonlethal *ts* mutants in *recA*, *recB*, *recC*, or *polA* (83).

It has been suggested that DNA methylation enables discrimination between parental (methylated) and nascent undermethylated strands in postreplication mismatch repair (84, 85). Direct evidence supporting this suggestion was obtained by using transfection assays with heteroduplex lambda DNA in which one strand was methylated and the other not, each strand carrying different genetic markers. Only the markers on the methylated strand were expressed (86). This mechanism enables an efficient repair system to ensure repair of a mismatched base exclusively on the newly synthesized strand.

A meiotic recombination model that predicts hemimethylated DNA in germ cells has been proposed (87). The DNA of sperm was actually found to be differently methylated than somatic DNA (45, 63, 88), but more experiments are needed to test the validity of this model.

VI. Conclusions and Prospects

All sites in bacterial DNA that can potentially be methylated are indeed fully methylated (16), but only some of the methylatable sites in eukaryotic DNA are methylated (10). This difference in distribution of methyl groups reflects, most probably, different mechanisms of methylation. While the pattern of methylation of prokaryotic DNA can be determined exclusively by the availability of methylatable sites, the distribution of methyl groups in eukaryotic DNA might reflect a more complex mechanism. One element in the determination of the pattern of DNA methylation might be the nucleosomal structure of eukaryotic chromatin. As a first step to examine this possibility, the sequence of events—DNA replication, DNA methylation, and nucleosome assembly—should be established. If nucleosome formation precedes DNA methylation, it might affect the distribution of methyl groups. This and other questions with regard to the mechanism of the DNA methylation process must be answered before the function of DNA methylation can be fully understood.

Slow progress is being made with respect to elucidation of the biological functions played by the methyl groups in DNA. Several observations suggest a relationship between DNA methylation and gene expression. Studies on the pattern of methylation of the globin gene (26, 65), ovalbumin gene (45), and several viral genes (29, 30) revealed that, in general, the respective gene is undermethylated at specific sites in the tissue where the gene is expressed. The question whether this undermethylation is required for gene expression, or alternatively, is a result of the transcriptional process, remains to be resolved.

Tissue specificity with respect to the pattern of methylation can be observed when specific methylatable sites in single genes are studied. However, no significant differences are noted when the average methylation of DNA from various tissues is studied. Therefore, in order to study the possible involvement of DNA methylation in differentiation and regulation of gene expression, attention should be focused on specific sites.

Preliminary results of experiments in bacteria demonstrate that methylase-deficient mutants are hypersensitive to mutagenesis and have higher rates of DNA recombination (18). There are also preliminary indications that DNA methylation is coupled to DNA replication in *E. coli* (74). These experiments should be extended to eukaryotic systems.

Although there is almost no evidence for the existence of restriction enzymes in eukaryotic cells, it should not discourage continuation of research aimed at the elucidation of restriction-modification systems in eukaryotes.

Acknowledgments

We are grateful to Drs. H. Cedar and S. Ben-Sasson for critical reading of the manuscript. Experimental work done in our laboratory and described here has been supported by National Institutes of Health Grant GM20483.

References

1. R. D. Hotchkiss, *JBC* **175**, 315 (1948).
2. D. B. Dunn and J. D. Smith, *BJ* **68**, 627 (1958).
3. M. Gold, J. Hurwitz and M. Andres, *PNAS* **50**, 164 (1963).
4. J. Doskočil and Z. Šormovà, *BBA* **95**, 513 (1965).

5. G. R. Wyatt, *BJ* **48**, 584 (1951).
6. A. D. Riggs. *Cytogenet. Cell Genet.* **14**, 9 (1975).
7. R. Holliday and J. E. Pugh, *Science* **187**, 226 (1975).
8. R. Sager and R. Kitchin, *Science* **189**, 426 (1975).
9. H. J. Taylor, in "Molecular Genetics" (J. H. Taylor, ed.), Vol. 3, p. 89. Academic Press, New York, 1979.
10. A. Razin and A. D. Riggs, *Science* **210**, 604.
11. H. O. Smith, *Science* **205**, 455 (1979).
12. A. Razin, in "The Single-Stranded DNA Phages" (D. T. Denhardt, D. Dressler, and D. S. Ray, eds.), Monogr. Ser., p. 165. Cold Spring Harbor Lab., Cold Spring Harbor, New York, 1978.
13. V. F. Nesterenko, Ya. I. Buryanov and A. A. Baev, *Biokhimiya* **44**, 130 (1979).
14. D. Simon, F. Grunert, U. v. Acken, H. P. Döring and H. Kröger, *NARes* **5**, 2153 (1978).
15. R. L. P. Adams, E. L. McKay, L. M. Craig and R. H. Burdon, *BBA* **561**, 345 (1979).
16. A. Razin, S. Urieli, Y. Pollack, Y. Gruenbaum and G. Glaser, *NARes* **8**, 1783 (1980).
16a. Y. Pollack, R. Stein, A. Razin and H. Cedar, *PNAS* **77**, 6463 (1980).
17. R. J. Roberts, *NARes* **8**, r63 (1980).
18. M. G. Marinus and N. R. Morris, *J. Bact.* **114**, 1143 (1973).
19. S. Hattman, *J. Bact.* **129**, 1330 (1977).
20. S. Urieli, G. Glaser, J. Mager and A. Razin, *Isr. J. Med. Sci.* **15**, 792 (1979).
21. I. I. Nikolskaya, N. G. Lopatina, N. V. Anikeicheva and S. S. Debov, *NARes* **7**, 517 (1979).
22. P. M. M. Rae and R. E. Steele, *BioSystems* **10**, 37 (1978).
23. J. Singer, J. Robert-Ems and A. D. Riggs, *Science* **203**, 1019 (1979).
24. J. Singer, J. Robert-Ems, F. W. Luthardt and A. D. Riggs, *NARes* **7**, 2369 (1979).
25. Y. Gruenbaum and A. Razin, unpublished results.
26. L. H. T. van der Ploeg and R. A. Flavell, *Cell* **19**, 947 (1980).
27. H. Cedar, A. Solage, G. Glaser and A. Razin, *NARes* **6**, 2125 (1979).
28. I. B. Dawid, D. D. Brown and R. H. Reeder, *JMB* **51**, 341 (1970).
29. R. C. Desrosiers, C. Mulder and B. Fleckenstein, *PNAS* **76**, 3839 (1979).
30. D. Sutter and W. Doerfler, *PNAS* **77**, 253 (1980).
31. M. Meijer, E. Beck, F. G. Hansen, H. E. N. Bergmans, W. Messer, K. von Meyenberg and H. Schaller, *PNAS* **76**, 580 (1979).
32. K. Sugimoto, A. Oka, H. Sugisaki, M. Takanami, A. Nishimura, Y. Yasuda and Y. Hirota, *PNAS* **76**, 575 (1979).
33. F. Sanger, A. R. Coulson, T. Friedmann, G. M. Air, B. C. Barrell, N. L. Brown, J. C. Fiddes, C. A. Hutchison, III, P. M. Solocombe and M. Smith, *JMB* **125**, 225 (1978).
34. A. Razin, J. W. Sedat and R. L. Sinsheimer, *JMB* **53**, 251 (1970).
35. A. S. Lee and R. L. Sinsheimer, *J. Virol.* **14**, 872 (1974).
36. R. Salomon, A. M. Kaye and M. Herzberg, *JMB* **43**, 581 (1969).
37. O. J. Miller, W. Schnedl, J. Allen and B. F. Erlanger, *Nature* **251**, 636 (1974).
38. K. Harbers, B. Harbers and J. H. Spencer, *BBRC* **66**, 738 (1975).
39. A. Solage and H. Cedar, *Bchem* **17**, 2934 (1978).
40. J. Singer, R. H. Stellwagen, J. Roberts-Ems and A. D. Riggs, *JBC* **252**, 5509 (1977).
41. A. P. Bird, M. H. Taggart and B. A. Smith, *Cell* **17**, 889 (1979).
42. A. Razin and H. Cedar, *PNAS* **74**, 2725 (1977).
43. H. Weintraub, *NARes* **7**, 781 (1979).
44. E. M. Southern, *JMB* **98**, 503 (1975).
45. J. L. Mandel and P. Chambon, *NARes* **7**, 2081 (1979).
46. D. Billen, *JMB* **31**, 477 (1968).

47. C. Lark, *JMB* **31**, 389 (1968).
48. J. M. Pollock, Jr., M. Swihart and J. H. Taylor, *NARes* **5**, 4855 (1978).
49. A. P. Bird, *JMB* **118**, 49 (1978).
50. D. Drahovsky and N. R. Morris, *JMB* **57**, 475 (1971).
51. A. P. Bird and E. M. Southern, *JMB* **118**, 27 (1978).
52. E. F. Fisher and M. H. Caruthers, *NARes* **7**, 401 (1979).
53. E. Scarano, *Adv. Cytopharmacol.* **1**, 13 (1971).
54. E. Scarano, M. Iaccarino, P. Grippo and E. Parisi, *PNAS* **57**, 1394 (1967).
55. P. Grippo, E. Parisi, C. Carestia and E. Scarano, *Bchem* **9**, 2605 (1970).
56. T. W. Sneider and V. R. Potter, *JMB* **42**, 271 (1969).
57. T. W. Sneider, *JMB* **79**, 731 (1973).
58. A. P. Bird, *NARes* **8**, 1499 (1980).
59. H. D. Royer and R. Sager, *PNAS* **76**, 5794 (1979).
60. J. W. Kappler, *J. Cell. Physiol.* **78**, 33 (1971).
61. B. F. Vanyushin, A. L. Mazin, V. K. Vasilyev and A. N. Belozersky, *BBA* **299**, 397 (1973).
62. B. F. Vanyushin, S. G. Tkacheva and A. N. Belozersky, *Nature* **225**, 948 (1970).
63. C. Waalwijk and R. A. Flavell, *NARes* **5**, 4631 (1978).
64. C. Shen and T. Maniatis, unpublished.
65. J. D. McGhee and G. D. Ginder, *Nature* **280**, 419 (1979).
66. M. Wigler, R. Sweet, G. K. Sim, B. Wold, A. Pellicer, E. Lacy, T. Maniatis, S. Silverstein and R. Axel, *Cell* **16**, 777 (1979).
67. S. M. Taylor and P. A. Jones, *Cell* **17**, 771 (1979).
68. P. A. Jones and S. M. Taylor, *Cell* **20**, 85 (1980).
69. J. K. Christman, P. Price, L. Pedriman and G. Acs, *EJB* **81**, 53 (1977).
70. M. Terada, H. Fujiki, P. A. Marks and T. Sugimura, *PNAS* **76**, 6411 (1979).
71. W. Arber, This Series **14**, 1 (1974).
72. S. Lacks and B. Greenberg, *JBC* **250**, 4060 (1975).
73. R. H. Burdon, A. C. B. Cato and I. S. Valladsen, in "Gene Functions" (S. Rosenthal *et al.*, eds.), FEBS 12th Meet., Vol. 51, p. 159. Elsevier, Amsterdam, 1978.
74. M. Szyf, N. Galili and A. Razin, to be published.
75. A. Razin, J. W. Sedat and R. L. Sinsheimer, *JMB* **78**, 417 (1973).
76. A. Razin, D. Goren and J. Friedman, *NARes* **2**, 1967 (1975).
77. J. Friedman and A. Razin, *NARes* **3**, 2665 (1976).
78. J. Friedman, A. Friedmann and A. Razin, *NARes* **4**, 3483 (1977).
79. J. H. Taylor, *Chromosoma* **62**, 291 (1977).
80. C. Coulondre, J. H. Miller, P. J. Farabough and W. Gilbert, *Nature* **274**, 775 (1978).
81. T. Lindahl, This Series **22**, 135 (1979).
82. W. Salser, *CSHSQB* **42**, 985 (1977).
83. M. G. Marinus and N. R. Morris, *JMB* **85**, 309 (1974).
84. E. B. Konrad, *J. Bact.* **130**, 167 (1977).
85. R. Wagner, Jr. and M. Meselson, *PNAS* **73**, 4135 (1976).
86. B. W. Glickman, P. van den Elsen and M. Radman, *Mol. Gen. Genet.* **163**, 307 (1978).
87. T. W. Sneider, J. Kaput, D. Neiman and B. Westmoreland, in "Transmethylation" (E. Usdin, R. T. Borchardt and C. R. Creveling, eds.), p. 473. Elsevier, Amsterdam, 1979.
88. J. Kaput and T. W. Sneider, *NARes* **7**, 2303 (1979).
89. Y. Gruenbaum, R. Stein, H. Cedar and A. Razin, to be published.
90. H. Cedar, R. Stein, Y. Pollack and A. Razin, to be published.

Mechanisms of DNA Replication and Mutagenesis in Ultraviolet-Irradiated Bacteria and Mammalian Cells

JENNIFER D. HALL AND
DAVID W. MOUNT

*Department of Cellular and
Developmental Biology,
College of Liberal Arts, and
Department of Microbiology,
College of Medicine,
University of Arizona,
Tucson, Arizona*

I. Introduction	54
A. Introductory Comments and Scope of the Review	54
B. Overview of DNA Repair Mechanisms	56
C. Repair-Deficient Mutants of Mammalian Cells and Bacteria	58
II. DNA Synthesis in Ultraviolet-Irradiated Bacteria	60
A. Inhibition of DNA Synthesis in Bacteria by UV Irradiation	60
B. Gaps in Replicated DNA	60
C. Evidence for Gaps Opposite Pyrimidine Dimers	61
D. Recombinational Mechanism for Removal of Gaps	63
E. Recovery of the Ability to Synthesize High-Molecular-Weight DNA	66
F. Evidence for Continuous DNA Synthesis Past Pyrimidine Dimers *in Vivo*	67
G. Effects on Initiation of Chromosome Replication	69
H. Model for Role of *recA* and *lexA* Genes in DNA Repair	70
III. DNA Synthesis in Ultraviolet-Irradiated Mammalian Cells	75
A. Control of DNA Replication	75
B. Effects on the Cell Cycle	77
C. Inhibition of DNA Synthesis	77
D. Inhibition of Replicon Initiation	79
E. Inhibition of DNA Elongation	81
F. Evidence for Blockage of DNA Synthesis at Pyrimidine Dimers	82
G. Replication Past Sites of UV Damage	89
H. Do Gaps Opposite Pyrimidine Dimers Exist in Replicated DNA?	90
I. Does Genetic Recombination Play a Role in DNA Repair or in Mechanisms of Tolerance of Ultraviolet Damage?	92
J. Mechanisms for Replication of Damaged DNA	96

K. Evidence for and against the Existence of an Inducible Mechanism for Replication of Damaged DNA 98
IV. Mutagenesis by Ultraviolet Radiation 100
   A. Mutagenesis in Bacteria 100
   B. Mutagenesis in Mammalian Cells 106
V. Mechanism for Reactivation of Ultraviolet-Damaged Viruses 108
   A. Repair of Bacteriophage 108
   B. Mutagenesis of Ultraviolet-Irradiated Bacteriophage 110
   C. Repair of Animal Viruses 112
   D. Mutagenesis Associated with Ultraviolet Reactivation of Animal Viruses 114
VI. Effects of Ultraviolet Irradiation on DNA Synthesis *in Vitro* 116
   A. Replication of Ultraviolet-Damaged DNA 117
   B. Methods for Measurement of Replication Fidelity 118
VII. Summary and Future Perspectives 121
   References 122
   Note Added in Proof 126

I. Introduction

A. Introductory Comments and Scope of the Review

Several developments in the past decade have aroused much interest in DNA repair. First, there is evidence for a close relationship between DNA repair, mutagenesis, and carcinogenesis, as revealed by the observation that many carcinogenic chemicals cause DNA damage and induce mutations in bacteria and mammalian cells. Second, it has been speculated that DNA repair is related to biological aging. For example, it has been proposed that a genetically programmed decline in repair ability could lead to cellular accumulation of mutated nucleic acids and dysfunctional proteins, which could thus account for many of the characteristics of aging (*1, 2*). Third, a number of human genetic disorders involving deficiencies in DNA repair have been identified. Many of these disorders are associated with developmental abnormalities, such as neurological or immunological dysfunction. Therefore, cellular functions defective in individuals with these diseases appear to be required for both DNA repair and normal development, suggesting a close relationship between these biological processes.

The promise of so much valuable information in the study of DNA repair has not yet been fully achieved. Much research in this field is highly descriptive and has used relatively few biochemical techniques to study a wide variety of cell types or types of DNA damage. Consequently, one purpose of this review is to advocate the development of new biochemical technology for the study of DNA repair.

Several reviews on the general subject of DNA repair have recently been published; those by Hanawalt et al. (3) and Cleaver (4) are the most current. Related reviews include discussions of photoreactivation (5); enzymology of excision repair (6, 7); repair of DNA damaged by carcinogenic chemicals (8); repair of DNA damaged by far-ultraviolet light (9); use of enzymes that recognize damaged lesions to monitor DNA repair (10); human disorders with deficiencies in DNA repair (11); the relationship of repair deficiencies to cancer (12); mutagenesis in bacteria (13) and in mammalian cells (14); analysis of the fidelity of DNA repair (15); the concepts of multiple repair pathways (16) and inducible, error-prone repair (17) in *Escherichia coli*; mechanisms of genetic recombination (18); and mechanisms of repair in bacteriophages (19) and in yeast (20), the latter as a model system for analysis of DNA repair in eukaryotes. An earlier review of the subject material covered in the present review is that by Cleaver (21). Three other reviews have dealt specifically with the problems of adapting bacterial models to mammalian cells (4, 22, 23), as we do here.

The major goal of this review is to evaluate critically studies on DNA replication in bacteria and mammalian cells irradiated by ultraviolet (UV) light. This topic is of interest because cellular mechanisms for replicating damaged DNA are functionally related to cellular control of UV-light-induced mutagenesis. We have chosen to restrict our analysis to UV damage, since a lesion in DNA that results in major biological and mutageneic effects, the pyrimidine dimer, has been identified. Other pathways for repair of pyrimidine dimers (e.g., excision repair and photoreactivation) have been extensively reviewed previously (3, 5-7) and will be mentioned here only briefly.

In particular, we discuss the effects of UV irradiation on DNA synthesis in bacteria, mammalian cells, bacteriophages, and animal viruses. Possible mechanisms for replication of damaged chromosomes are described, and the mechanisms of UV-induced mutagenesis are analyzed. Because studies performed in bacteria have frequently served as models for similar processes in mammalian cells, the bacterial studies are reviewed first. The extent to which these models apply to mammalian cell systems is discussed, and, where appropriate, new models are presented to explain data from mammalian cells that fail to be explained satisfactorily by the bacterial models.

Finally, we have sought to write a critical review, not a literature survey. The papers chosen for discussion were selected because they present new or different views of the subject. A more complete exposure to the areas discussed may be obtained from the review articles listed above.

B. Overview of DNA Repair Mechanisms

The major photoproduct induced by irradiation of both single- and double-stranded DNA by 254 nm light is the intrastrand pyrimidine dimer[1] (24, 25). It is generally assumed that this product is largely responsible for the biological effects of UV irradiation because these effects can be reversed by photoreactivation, an enzymic process that monomerizes the dimers (5).

Pyrimidine dimers can be removed from cellular DNA by two major mechanisms in bacteria and mammalian cells: photoreactivation (5) and excision repair (3). Those pyrimidine dimers not removed by a repair mechanism present an obstacle to DNA replication, as they distort the DNA backbone so that normal base-pairing cannot occur opposite the dimer sites. Consequently, unless there is some mechanism permitting synthesis past these damaged lesions, synthesis would be permanently blocked there. After replication of irradiated DNA, the remaining DNA damage can be enzymically removed by one of the mechanisms listed above, or may instead be tolerated by additional rounds of chromosomal replication (3).

Below we summarize briefly the current understanding of the photoreactivation and excision repair pathways, since these processes frequently interact with the mechanisms that allow replication of damaged DNA templates. Photoreactivation, illustrated in Fig. 1A, results from the action of a single cellular enzyme[2] (5), present both in bacteria and in mammalian cells. When UV-irradiated cells are exposed to visible light, the photoreactivation enzyme is activated to monomerize pyrimidine dimers directly, without removing them from the nucleic acid. Consequently, this process should not result in the production of mutations in the re-

---

[1] "Cyclobutadipyrimidine," in "Enzyme Nomenclature," 1978, is the approved generic term for "pyrimidine dimer." That produced from two thymines has the name (in *Chemical Abstracts*) hexahydro-4a,4b-dimethylcyclobuta[1,2-d; 4,3-d']dipyrimidine-2,4,5,7 ($3H,6H$)tetrone. [Ed.]

[2] EC 4.1.99.3, deoxyribodipyrimidine photolyase (Recommended Name), deoxyribocyclobutadipyrimidine pyrimidine-lyase (Systematic Name) *in* "Enzyme Nomenclature," 1978. [Ed.]

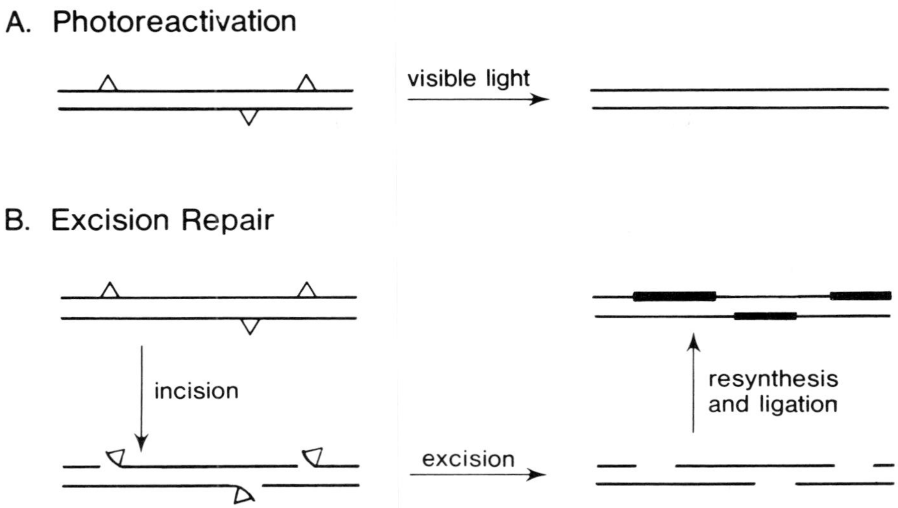

A. Photoreactivation

B. Excision Repair

FIG. 1. Repair pathways for removal of pyrimidine dimers from DNA. (A) Photoreactivation requires exposure of the photoreactivation enzyme to visible light and results in monomerization of the dimers by this activated enzyme *in situ*. (B) Excision repair involves cutting of the DNA adjacent to the dimers by an endonuclease, removal of single-stranded regions containing the dimers by an exonuclease, filling of the resulting gaps by DNA synthesis, and ligation of the repaired DNA to form continuous DNA molecules.

paired DNA. Since the photoreactivation pathway acts specifically to repair pyrimidine dimers, it is frequently used in experimental procedures to remove pyrimidine dimers from cells in which DNA repair is being studied.

A proposed scheme for excision repair (Fig. 1B) involves removal of a single-stranded segment of DNA containing the damage, followed by repair synthesis to fill the gap. The mechanism requires at least four enzyme activities: (*a*) a specific endonuclease, which makes a cut adjacent to the dimer in the DNA strand containing the dimer [see Note Added in Proof (a)]; (*b*) a second nuclease activity (probably an exonuclease), which removes the region containing the dimer; (*c*) a polymerase to synthesize DNA in the excised region; and (*d*) a ligase to restore the final phosphodiester bond. While the above enzyme activities of bacterial cells have been well studied, the comparable mammalian enzymes have been relatively poorly characterized (3, 6, 7).

In *Escherichia coli*, damaged regions of DNA are subjected to phosphodiester hydrolysis by a complex of three proteins, the *uvrA*, *uvrB*, and *uvrC* gene-products (3, 7). This complex has been purified and partially characterized. Other endonucleases that cleave UV-irradiated DNA specifically have also been isolated from *Micrococcus luteus* (6, 10) and from T4 bacteriophage-infected *E. coli* (T4 endonuclease V; 6, 7, 10). These latter enzymes have been used to monitor the presence of pyrimidine dimers in UV-irradiated DNA *in vitro* (10). In mammalian cells, identification of an enzyme from calf thymus that cuts DNA containing pyrimidine dimers has recently been reported (27). Other endonucleases from mammalian cells that cut UV-irradiated DNA appear to recognize photoproducts other than pyrimidine dimers (28, 29).[3]

Genetic and biochemical studies have suggested that *E. coli* might utilize at least three enzymes for excision of the damaged regions from irradiated DNA: the 5'-exonuclease of polymerase I (3, 30), the 5'-exonuclease of polymerase III (3, 31), and the single-strand-specific exonuclease VII (3, 32). The fact that two of these enzymes are associated with polymerase activities has suggested that the excision and resynthesis steps in this repair pathway might occur simultaneously, possibly by a "nick-translation" mechanism.

There are reports of mammalian exonucleases capable of excising pyrimidine dimers *in vitro* from DNA previously incised by UV-specific endonucleases (33, 34). However, whether these exonucleases function as excision enzymes *in vivo* has not yet been determined. In addition, while the mammalian DNA polymerase β is currently thought to be involved in DNA repair synthesis because its level of activity does not increase during the DNA synthetic (S) phase of the cell cycle (35), it is not known whether this polymerase performs the resynthesis step in excision repair.

C. Repair-Deficient Mutants of Mammalian Cells and Bacteria

Two major classes of UV-sensitive mutants of *E. coli* have been characterized: mutants unable to perform excision repair; and mutants deficient in replication of damaged DNA. A mutant deficient in both these processes ($uvrA^-$ $recA^-$) is killed by the presence of only one or two pyrimidine dimers per chromosome, suggesting that a single unexcised dimer can be lethal (36). Consequently, the ability of cells to replicate damaged DNA is important for cell sur-

---

[3] The nomenclature, including abbreviations and symbols, of a wide variety of pyrimidine photoproducts is presented in Cohn *et al.* (29a). [Ed.]

vival. [A complete list of *E. coli* mutants deficient in DNA repair has been prepared by Hanawalt *et al.* (3).]

Excision-deficient (uvr^-) mutants of *E. coli* have been used extensively to study the replication of DNA in UV-damaged bacteria (37–39), since pyrimidine dimers are not removed from the DNA in these mutants except by photoreactivation (photolysis). Therefore, these cells must contend with the difficulties of synthesis past pyrimidine dimers at each round of replication.

Among the UV-sensitive mutants that show deficiencies in replication of UV-damaged DNA are $lexA^-$, $recA^-$, and $recF^-$ mutants of *E. coli* (40–42). The $recA^-$ mutants are defective in genetic recombination, as indicated by the inability of a $recA^-$ recipient to integrate DNA transferred from an Hfr donor (43). The $recF^-$ mutants show a similar deficiency when present in another genetic background ($recB^-$ $sbcB^-$ or $recC^-$ $sbcB^-$) (43). This association between deficiencies in DNA repair and in recombination suggests that genetic recombination is important in replication of damaged DNA. The $lexA^-$ mutants are not recombination deficient, but appear instead to be deficient in regulation of the *recA* gene (44, 45; see Section II,H).

Among those mammalian cell lines exhibiting abnormal sensitivity to UV are those derived from human patients with the genetic disease xeroderma pigmentosum (XP). Most XP cells are deficient in the ability to excise pyrimidine dimers from their DNA and, consequently, appear to be similar to the *uvr* mutants of *E. coli* (11, 12). By comparison with wild-type cells, they show reduced cellular survival following UV irradiation, and reduced repair of UV-irradiated animal viruses (46–49). When cells from different XP patients are fused into heterokaryons, normal excision repair can occur by complementation (50). By this technique, seven complementation groups for XP have been found (50–52). The biochemical deficiency in these cells has not been identified, but it is thought they carry a defect in the incision step of excision repair, as they exhibit low levels of excision-induced, single-strand breaks in cellular DNA following UV irradiation (53). The incision step in dimer excision may, therefore, require the sequential actions of several gene products or, alternatively, the activity of a multienzyme complex.

A second class of XP cells, XP variants, exhibit normal level of excision repair synthesis, repair UV-irradiated animal viruses at near normal levels (46, 47, 54), and are only slightly sensitive to UV (55). However, patients from whom these cells were derived exhibit the full severity of the disease, including hypersensitivity of skin to sunlight, abnormal skin pigmentation, and excessive production of skin tumors (*11, 12*). Consequently, XP variant cells appear to carry deficiencies that substantially affect the response of these cells to UV. Although these deficiencies are not fully understood, XP variant cells appear to be defective in replication of DNA following UV irradiation (*56, 57*). In addition, they appear to repair single-strand breaks, including those created by excision repair, more slowly than do normal cells (*53, 58*). Other properties of these mutant cell lines are discussed in the relevant section of this review (Section IV,B).

II. DNA Synthesis in Ultraviolet-Irradiated Bacteria

A. Inhibition of DNA Synthesis in Bacteria by UV Irradiation

Treatment of bacteria with UV inhibits DNA synthesis, and the extent of inhibition increases with exposure dose (*37*). Photoreactivation largely relieves this inhibition, showing that the inhibition is caused primarily by pyrimidine dimers (*59*). The extent of inhibition from a given dose is greater in excision-deficient strains than in proficient strains, suggesting that normal rates of DNA synthesis can be restored following excision of pyrimidine dimers (*60*).

B. Gaps in Replicated DNA

A detailed study of DNA synthesis in excision-deficient ($uvrA^-$) bacteria (*37*) suggests that these cells synthesize DNA past sites of pyrimidine dimers following UV irradiation. When cells were pulse-labeled with [^{3}H]thymidine shortly after irradiation, the average molecular weight of the newly synthesized DNA was much lower that that of DNA from unirradiated cells. Three mechanisms that might account for this low-molecular-weight DNA were considered (*37*): (i) a reduced rate of DNA synthesis; (ii) induction of new rounds of DNA synthesis from the origin of replication; or (iii) the existence of discontinuities not found in DNA synthesized in unirradiated cells.

A reduced rate of DNA synthesis in irradiated cells (mechanism i) at first appeared reasonable, since, for an equivalent period of labeling, the total amount of DNA synthesized in irradiated cells was less than that in unirradiated cells. Also, it was observed (*37*) that shortening the labeling period for unirradiated cells produced shorter DNAs than that found after longer pulses. However, when the labeling periods for irradiated and unirradiated cells were adjusted to allow equivalent levels of incorporation in both cell cultures, the average size of the DNA synthesized in irradiated cells was still considerably less than that of DNA from unirradiated cells (37). Therefore, the presence of short DNA molecules made in irradiated cells is not explained by a reduced rate of DNA synthesis.

Under certain conditions, UV irradiation induces a new cycle of chromosomal replication from the normal origin of *E. coli* (61) (see Section II,G). To determine whether reinitiation (mechanism ii) might produce short DNA strands in irradiated cells, Rupp and Howard-Flanders (37) incubated cultures in unlabeled medium for 50 minutes following UV irradiation, a period sufficient to complete any newly induced rounds of replication. The cells were then pulse-labeled. The DNA synthesized under these conditions was also of low molecular weight, indicating that most of the low-molecular-weight DNA made in irradiated cells does not arise by reinitiation of synthesis.

Several observations are consistent with the presence of discontinuities in DNA synthesized in irradiated *E. coli* (mechanism iii). First, the average size of nascent DNA molecules is inversely proportional to the UV dose and is approximately equal to the average distance between adjacent pyrimidine dimers in the same DNA strand (37). Second, photoreactivation of irradiated cells prior to labeling increases the size of the DNA (62). Third, excision-proficient strains recover the ability to synthesize large DNAs more rapidly than deficient strains, presumably because dimer excision restores normal DNA synthesis (62). Finally, the structure of newly synthesized DNA from irradiated cells has been analyzed by chromatography on benzoylated, naphthoylated DEAE-cellulose. While this analysis does not reveal the precise molecular structure of this DNA, the results are consistent with a structure having the properties of double-stranded DNA containing single-stranded regions approximately 1000 nucleotides long (63). These results strongly suggest that there are gaps in newly replicated DNA strands in irradiated cells, and these gaps may account for the observation of smaller DNAs in alkaline sucrose gradients.

C. Evidence for Gaps Opposite Pyrimidine Dimers

As described above, the length of newly synthesized DNA molecules from irradiated cells approximates the average distance between adjacent pyrimidine dimers in the same DNA strand (37). This correlation suggests that DNA synthesis might be blocked at pyrimidine dimers, but might then reinitiate downstream from these UV lesions, leaving gaps opposite the dimers. This mechanism seems reasonable in view of the structural deformity created in DNA by the dimer. This deformity moves the joined pyrimidine bases out of position, so that proper hydrogen bonding with incoming dNTPs is difficult, if not impossible, to achieve.

Evidence favoring the presence of gaps opposite dimers in *E. coli* comes from studies of conjugation crosses between irradiated donors carrying sex factor DNA and unirradiated recipient cells (64). In these experiments, the survival of a chromosomal marker on the sex factor DNA was measured after transfer to a recipient cell. During conjugation crosses with unirradiated donor cells, one of the two sex-factor DNA strands is transferred to the recipient cell, and is replicated in the recipient during transfer (65, 66). A similar process is presumed to occur when DNA is transferred from an irradiated donor (67). Consequently, DNA molecules transferred from an irradiated strain may consist of one donor strand containing dimers and one newly replicated strand containing gaps. Excision repair of such a structure would not be expected to occur if these gaps were opposite dimers, since there would be no template in the strand opposite the dimer. In the above experiment (64), for a given exposure dose, the survivals of the chromosomal marker carried by the irradiated sex factor were similar in excision-deficient and proficient recipient cells. It was concluded that pyrimidine dimers in the transferred episomes could not be excised and that gaps in the newly made strands were opposite dimers (64).

However, the complexity of the biological system in these experiments (64) makes it difficult to rule out other explanations. As discussed in Section II,D, partially replicated sex-factor DNA carrying a chromosomal marker can undergo extensive genetic recombination with homologous DNA in the recipient chromosome. These recombination events provide a mechanism for recovery of the chromosomal marker carried on the irradiated sex factor. As a consequence, if excision repair of sex factor DNA is possible, it might go undetected in these experiments, since recombination may provide a more efficient mechanism for marker recovery than excision repair. In addition, the transferred DNA causes induction of *recA* protein, and large quantities of this protein may induce mutagenesis of the sex factor (see Section IV,A). This mutagenesis might influence survival of the sex factor or the scored marker in ways not predictable without a much more rigorous analysis of this system. For example, sex-factor DNA might survive, but might not be scored owing to a mutation in the selected marker.

To summarize, newly transferred DNA from irradiated donor

cells may be discontinuous. Whether the gaps are opposite pyrimidine dimers depends on the interpretation of the sex-factor experiment described above. The most convincing argument for this dimer-gap structure is that the structural deformity created in the template DNA strand by the pyrimidine dimer should make effective hydrogen bonding with incoming nucleotides difficult, if not impossible.

D. Recombinational Mechanism for Removal of Gaps

Pulse-labeled DNA from irradiated cells, which initially appears as short (low-molecular-weight) strands, eventually becomes covalently associated with much longer DNA strands (37) after further incubation of these cells in unlabeled medium. This observation suggests that the shorter strands are elongated by some mechanism that eliminates the gaps present before the chase period (37). This elongation process, referred to as "postreplication repair," is not a repair process removing DNA damage, such as occurs in excision repair and photoreactivation, but rather a process enabling cells to replicate damaged chromosomes. Possible mechanisms for postreplication repair in bacteria are (i) filling of gaps in nascent strands of a damaged chromosome by homologous DNA from the sister chromosome, and (ii) appearance of a mechanism for replication past sites of pyrimidine dimers, e.g., by modification of either the DNA replication machinery or the structure of the partially replicated DNA molecules so that bases can be inserted opposite pyrimidine dimers.

Both genetic and biophysical data suggest that filling of gaps in nascent DNA from irradiated bacteria can occur by a recombination mechanism (mechanism i). First, *E. coli* mutants deficient in genetic recombination (*recA$^-$*) can synthesize short DNA strands following irradiation, but fail to elongate this DNA during subsequent incubation (62). Second, UV irradiation of *E. coli* stimulates the frequency of DNA exchanges detected either by genetic crosses or by biophysical techniques (38, 64, 68, 69).

Experiments with bacteriophage lambda (69) suggest that pyrimidine dimers in phage DNA stimulate genetic exchanges with homologous phage chromosomes and that chromosomes that contain dimers must be replicated in order for such stimulation to occur. In these studies, *E. coli* excision-deficient (*uvrA$^-$*) lysogens containing a mutant prophage were superinfected with a second lambda mutant homologous to the prophage. Host and viral genotypes were varied so that replication of the incoming virus was either permitted or not permitted. During infections in which viral replication could occur, more recombinant bacteriophage was recovered when the superinfecting virus had been irradiated with UV than when unirradiated phage was employed. Furthermore, the yield of recombinant progeny was proportional to the UV dose. Pyrimidine dimers appeared to be required for the increased production of recombinants, because the stimulation was reduced (a) in a host able to excise pyrimidine dimers, and (b) in cells treated with photoreactivating light following infection.

Similar crosses (69) were carried out with superinfecting virus that could not replicate under the conditions of the experiment. The infected cells were not killed by this virus and could be grown for ten or more generations. During this time, the superinfecting virus genomes were diluted to the point that they were present in only a very small fraction of the progeny cells. The prophages in these cultures were then induced to replicate, and the number of recombinants among the progeny virus was measured. The frequency of recombinant phage recovered from cultures infected with unirradiated virus was very low and was stimulated to a small extent by UV irradiation of the superinfecting phage. However, this stimulation was presumably due to some unidentified photoproduct other than pyrimidine dimers in the irradiated DNA, since it was not influenced by the excision repair capability of the host nor decreased by a photoreactivation treatment.

The above experiments (69) demonstrate that pyrimidine dimers can stimulate recombination, but only during or after replication of damaged DNA. A possible explanation for this stimulation is that replication of irradiated DNA may form structures that can readily undergo recombination. However, equally plausible mechanisms are that (a) replication of the UV-damaged virus is blocked by a pyrimidine dimer and cannot continue until a genetic recombination event with the prophage has taken place, or (b) partially replicated, UV-damaged phage genomes induce host recombination functions that promote recombination between these genomes and the resident prophage.

Ultraviolet light also stimulates the frequency of genetic exchanges between cellular genes in *E. coli* (64, 68). In crosses between irradiated Hfr donors and unirradiated recipient cells, the recovery of recombinant progeny that had undergone multiple exchanges between donor DNA and the recipient chromosomes is substantially higher than in similar crosses with undamaged donors (64, 68). Single-stranded DNA transferred from an irradiated Hfr strain becomes double-stranded in the recipient cell (67). Consequently, it was proposed (64, 68) that the DNA strand transferred from the irradiated donor is replicated discontinuously and that the gaps stimulate recombination with the recipient chromosome. Similar studies (see Section II,C) have also been performed with an autonomously replicating sex factor carrying a chromosomal marker that can be incorporated into the recipient chromosome by a genetic recombination event (64). The frequency of such events also increases after irradiation of the donor strain, again possibly owing to gaps in the sex factor DNA opposite sites of pyrimidine dimers. However, there are alternative explanations for these results. For example, the dimer-gap structures produced by replication of UV-irradiated DNA might stimulate the production of *recA* protein (see Section II, H). High levels of this protein have been shown to promote multiple genetic exchanges, even in the absence of DNA damage (70).

A second method for detection of UV-induced exchanges is to examine the physical properties of DNA isolated from UV-irradiated cells for evidence of exchanges. Although this approach allows direct recovery of recombinant DNA, it is not known whether these DNA molecules are created by a repair pathway leading to cellular survival. To detect UV-induced DNA exchanges (38), excision-deficient ($uvrA^-$) *E. coli* cells were grown for several generations with ^{13}C and ^{15}N nutrients. The cells were then transferred for 20 minutes to medium containing normal isotopes, irradiated with UV, and finally incubated with radioactive thymidine. Analysis of the DNA in alkaline CsCl gradients indicated the presence of radioactive, single-stranded DNA of hybrid density, which might have been produced by several mechanisms: (a) end-labeling of heavy DNA strands by radioactive, light isotopes; (b) degradation of heavy isotopes and reincorporation into light DNA; or (c) recombination between fully heavy and fully light regions of the DNA. Hybrid molecules could not have arisen from end-labeling of heavy parental DNA strands by radioactive, light precursors, because the cells had been grown prior to irradiation in nonradioactive medium containing light isotopes for a period sufficient to complete any unfinished cycles of chromosomal replication. Nor could the hybrid molecules have arisen from degradation of DNA and reincorporation of the heavy level into progeny DNA. Such reutilization should lead to the production of progeny molecules uniformly labeled with both isotopes. The hybrid molecules detected were instead composed of discrete segments of nonradioactive, heavy DNA joined to segments of radioactive, light DNA. These segments could be separated by shearing the hybrid DNA and banding it in a CsCl density gradient. Therefore, the hybrid molecules appeared to have been produced by DNA exchanges. Inasmuch as such exchanges were not detected in unirradiated cells, they appear to have been induced by the UV damage. An estimate of the number of exchanges in this DNA indicated that a DNA segment containing both a heavy, nonradioactive and a light, radioactive region was formed for every 1.7 pyrimidine dimers induced by the irradiation treatment. Thus, there is approximately one genetic exchange for each pyrimidine dimer.

A third method to detect UV-induced genetic exchanges is to assess the presence of pyrimidine dimers in parental and newly synthesized DNA during growth after irradiation (39). Excision-deficient cells were used for these experiments to prevent loss of dimers by excision repair. Pyrimidine dimers were detected by treating DNA isolated from irradiated cells with the bacteriophage T4 endonuclease, which makes incisions next to pyrimidine dimers (endonuclease V). The size distribution of DNA treated with this endonuclease was compared to that of untreated DNA by sedimentation in alkaline sucrose gradients. The results showed that endonuclease-sensitive sites are gradually removed from parental DNA but simultaneously appear in the progeny DNA made after irradiation. This transfer of dimers suggests that genetic exchange events in irradiated cells can attach parental DNA containing pyrimidine dimers covalently to dimer-free progeny DNA. This attachment might occur during filling of gaps opposite sites of UV-induced damage in DNA replicated after irradiation.

The experiments described above suggest the following mechanism of DNA replication in UV-irradiated *E. coli*. Shortly after irradiation, the DNA replication complex is blocked at pyrimidine dimers on the bacterial chromosome. DNA synthesis then reinitiates on the other side of the dimer, approximately 1000 base-pairs down the strand. The effect of this termination and reinitiation sequence is to leave a gap in the newly synthesized strand extending from the site of the pyrimidine dimer to the reinitiation site. This gap is frequently filled by a recombination mechanism in which parental DNA from a sister chromosome is inserted into the gap. A model describing how these events might occur is discussed in Section II,H.

E. Recovery of the Ability to Synthesize High-Molecular-Weight DNA

When excision-deficient *E. coli* is irradiated with small doses of UV light, the overall rate of DNA synthesis is reduced shortly after irradiation, but recovers within 60 minutes (37, 38). Consequently, these cells appear able to replicate their DNA normally despite the continued presence of pyrimidine dimers in parental DNA. At the same later time, a significant fraction of the newly synthesized DNA sediments in alkaline sucrose gradients at the position of DNA from unirradiated cells rather than at the positions of the smaller DNA observed for nascent DNA isolated at earlier times after irradiation (38). Synthesis of these longer DNA molecules might be attributed to one or more of the following mechanisms: (a) synthesis occurs on dimer-free template strands; (b) DNA synthesis is still discontinuous, but the gaps are removed more rapidly than at earlier times; or (c) cells may have acquired the ability to replicate past pyrimidine dimers in a continuous fashion.

Experiments described in Section II,D suggest that pyrimidine dimers become equally distributed between parental and progeny DNA strands in excision-deficient cells as the time of incubation after irradiation is increased (39). This redistribution should result in a gradual dilution of pyrimidine dimers on each chromosome with every round of replication. However, the UV doses employed in the experiment described above (38) should have produced very large yields of dimers. Consequently, dimer-free templates for synthesis of high-molecular-weight DNA should not have been produced during the period of the experiment [mechanism (a) above].

As discussed previously (Sections II,B,D), newly replicated DNA synthesized at early times after irradiation appears abnormally small in alkaline sucrose gradients (37). However, some of the short molecules made at this early time eventually acquire the same (larger) size as molecules synthesized at later times. Ganesan (39) found that the high-molecular-weight DNA labeled 2 hours after irradiation contains just as many endonuclease-sensitive sites (i.e., pyrimidine dimers) as DNA molecules labeled immediately after irradiation and allowed to elongate for 2 hours. The appearance of pyrimidine dimers in newly synthesized DNA strands is thought to result from genetic recombination with parental, dimer-containing DNA, which fills gaps in the new strands. Consequently, the detection of pyrimidine dimers in nascent DNA made either at early or late times after irradiation suggests that gap-filling occurs at both these times. The larger DNA synthesized at late times might, therefore, result from more efficient gap filling, as proposed in mechanism (b) above.

F. Evidence for Continuous DNA Synthesis Past Pyrimidine Dimers *in Vivo*

Essentially all the larger DNA molecules detected in the above experiment contained pyrimidine dimers (39), a result apparently inconsistent with mechanism (c), synthesis past pyrimidine dimers, which should produce dimer-free molecules. However, the results described do not totally eliminate this latter mechanism. If mechanisms (*b*) and (*c*) were both functional at late times after UV irradiation, the production of dimer-free molecules would not be observed, despite continuous replication past damaged sites. It is also possible that dimer-free molecules might be synthesized, but dimers might be subsequently inserted into them by recombination.

In order to test more easily whether continuous replication past pyrimidine dimers might occur in *E. coli*, Caillet-Fauquet *et al.* (*71*) analyzed DNA synthesized on UV-irradiated, single-stranded bacteriophage ϕX174 templates following infection of both unirradiated cells and cells previously irradiated with a small dose of UV light. The irradiated cells were incubated, following UV exposure prior to infection, to allow expression of any inducible mechanism for replicating damaged DNA. The host cells were also nonpermissive for an amber mutation on the phage in gene A, which allows replication of the viral single-strands to double-stranded circles, but prevents further replication by a rolling-circle mechanism. At various times after infection, the phage DNA was extracted and analyzed in CsCl equilibrium density gradients, by hydroxyapatite chromatography, and by digestion with the single-strand specific nuclease, S1. These analyses indicated that prior UV irradiation of the virus inhibited the conversion of parental molecules to the progeny, double-stranded form. When previously irradiated host cells were infected with UV-irradiated virus, more double-stranded molecules were formed. Reinitiation of DNA synthesis from the origin of replication probably could not account for these results, because some viral molecules having the density and other properties of fully double-stranded molecules were produced in these infections. The enhanced replication of UV-damaged virus in irradiated cells was prevented by the addition of chloramphenicol to the irradiated cells before they were infected, demonstrating that protein synthesis in irradiated cells is necessary for enhancement.

While these results might be taken as evidence for enhanced synthesis of DNA past sites of pyrimidine dimers following UV irradiation of the host cells, another explanation must also be considered. Since these experiments were performed under conditions in which cells were infected with an average of three phage genomes per cell, viral recombination might account for the increased fraction of double-stranded molecules observed in irradiated cells. As discussed in Section II,H, UV irradiation of *E. coli* results in induction of the *recA* protein, and this protein might carry out recombination of partially replicated ϕX174 genomes. Consequently, the above experiment may not be evidence for direct insertion of nucleotides opposite sites of pyrimidine dimers. However, in Sections V,A and V,B we review other work demonstrating that increased survival and mutagenesis of UV-irradiated φX174 occurs in singly infected, UV-irradiated cells. These observations are consistent with a mechanism of DNA synthesis that is inducible and allows continuous synthesis of viral DNA past dimers. It is not presently known whether a similar mechanism acts on cellular DNA.

G. Effects on the Initiation of Chromosome Replication

Exponentially growing *E. coli* cells replicate their circular chromosomes bidirectionally, starting from a unique origin. After each replication cycle, protein synthesis is necessary to initiate a new cycle (*61, 72*). It has been proposed that one or more components of the replication complex become inactivated at the end of every round of replication, resulting in disassembly of the complex (reviewed in *61*). A new replication complex, which includes newly synthesized proteins, is then assembled. However, under certain conditions, e.g., UV irradiation or starvation of a thymidine-requiring mutant for thymidine, new cycles of replication can be initiated from the origin in the absence of protein synthesis.

Following thymidine starvation, subsequent addition of thymidine restores chromosomal replication at old replication forks, but new rounds of replication were also initiated (*61*). These new rounds of DNA synthesis occur semiconservatively and sequentially, starting at the origin, as in normally growing cells. The replication also requires the products of many of the *dna* genes, which are essential for normal DNA synthesis. In addition, this synthesis proceeds through formation of small replication intermediates (Okazaki fragments) and lasts for a time sufficient for several rounds of chromosomal replication.

This mode of chromosomal replication is also induced by UV light (*73*). Cultures of *E. coli* were irradiated with UV, grown for 40 minutes, then incubated in the presence of chloramphenicol to inhibit protein synthesis. DNA synthesis ceased in unirradiated, control cultures within 60 minutes after the addition of chloramphenicol, but continued in irradiated cultures for several hours. Protein synthesis was necessary during the 40-minute period after irradiation for this abnormal, UV-induced DNA replication. Replication was not observed in *lexA*$^-$ and *recA*$^-$ mutants of *E. coli*, suggesting that the abnormal DNA replication is one of a collective group of inducible functions, termed SOS functions, whose coordinate expression is thought to aid in cellular survival (see Section IV,A).

These studies suggest that, when DNA synthesis is interrupted by

DNA damage, a new mechanism of chromosomal replication that requires recA protein is induced in E. coli. This new type of DNA synthesis presumably aids in cellular survival and might be part of a general mechanism for recovery from DNA damage.

H. Model for Role of recA and lexA Genes in DNA Repair

The evidence presented in preceding sections shows that E. coli mutants deficient in excision repair can partially replicate UV-damaged DNA and subsequently reconstruct functional chromosomes from these partially replicated structures. Several different mechanisms seem to play a role in this process. In the period immediately following irradiation, progress of DNA replication forks along the bacterial chromosome appears to be blocked at each pyrimidine dimer, but new initiation events farther down the chromosome then occur, followed by synthesis to the next pyrimidine dimer. This pattern of discontinuous synthesis appears to arise from an inability of the replication complex to insert nucleotides opposite pyrimidine dimers. As the bacteria are incubated further, these interruptions are removed, probably by an efficient recombination mechanism that inserts homologous DNA derived from a sister chromosome into the gap. Some of the DNA strands synthesized at later times after UV irradiation are much longer, even though pyrimidine dimers are still abundant in the chromosome. A possible explanation for this efficient synthesis of continuous DNA strands at these times is that the recombination mechanism for filling discontinuities is so rapid that short DNA intermediate structures are not detectable by the pulse-labeling schedule employed. Another possibility is that changes either in the structure of the damaged chromosome, in the DNA replication complex itself, or in both of these elements could lead to the insertion of nucleotides opposite the pyrimidine dimer by the replication complex or by some enzymic mechanism.

Recent biochemical studies with bacterial mutants deficient in postreplication repair have led to a better understanding of the molecular mechanisms that may be involved in this process. We first describe the properties of some of these mutants and then discuss the characterization of some of the gene products necessary for the replication of damaged chromosomes.

1. INFLUENCE OF recA GENE ON DNA REPAIR

As already discussed (Section II,D), recA mutants of E. coli are deficient in genetic recombination (43) and are unable to repair gaps present in newly synthesized DNA in UV-irradiated cells (62). Since these mutants are extremely sensitive to UV, it appears that the *recA* gene product is essential for the reconstruction of functional chromosomes from damaged ones, and that failure to repair gaps in newly synthesized DNA of damaged cells leads to cell death. *recA⁻ uvr⁻* double mutants, deficient in both excision repair of pyrimidine dimers and postreplication repair, are killed with a dose of UV sufficient to induce approximately one or two pyrimidine dimers per chromosome (36). This sensitivity indicates that the *recA* protein is necessary for successful replication of a bacterial chromosome that contains pyrimidine dimers not removed by excision repair. The hypersensitivity of *uvrA⁻ recA⁻* mutants also demonstrates that *recA* mutations block all other modes of DNA repair in the cell except for excision repair.

recA protein may also be required for excision repair under circumstances where long stretches of DNA are removed exonucleolytically during dimer excision, leaving a long region in which repair synthesis must occur (3, 17). Such long regions may extend past dimers in the opposite strand. Successful refilling of these excision gaps, therefore, requires either a recombination event, insertion of bases opposite the dimers, or some other mechanism allowing formation of continuous DNA strands opposite dimers. Evidence that the *recA* gene product is involved in this process has been reviewed in detail elsewhere (3).

2. *recA* Promotes Pairing of Homologous DNA Strands

Howard-Flanders *et al.* (64) proposed that the ends of the DNA strands flanking the gaps in replicated, UV-damaged DNA promote pairing with homologous DNA in the sister chromosome. The pairing of homologous DNA in this manner has been considered to represent a possible initial step in recombination. This model is now supported by *in vitro* experiments utilizing highly purified *recA* protein, circular double-stranded DNA, and either homologous or nonhomologous single-stranded DNA fragments (74-78). The results suggest that single-stranded DNA with *recA* protein bound to it promotes transient denaturation of double-stranded DNA, and, if homology is present, this single-stranded DNA can invade the double-stranded molecule, displacing one of the strands to form a D-loop structure. If D-loop formations also occur *in vivo*, the D-loop might be acted upon by nucleases and DNA ligase to create a covalently joined molecule (78).

3. Regulation of *recA* Protein Synthesis

During normal growth in *E. coli*, a low basal level of *recA* protein is maintained. Following DNA damage or inhibition of DNA synthesis, *recA* protein is synthesized at a very high rate. Another *E. coli* gene, *lexA*, regulates this inducible expression of *recA*. Genetic studies with *lexA* and *recA* mutants suggest that the *lexA* product is a simple repressor of *recA* (e.g., see 45). As yet, there is no direct biochemical evidence that *lexA* is a repressor [see Note Added in Proof (b)]. However, *cis*-dominant mutations in the *recA* gene that allow constitutive *recA* protein synthesis, even in the absence of an inducing treatment, have been isolated (79, 80). These mutations may define an operator sequence to which the *lexA* protein may bind.

Following DNA-damaging treatments (or inhibition of DNA synthesis), *recA* protein becomes active as a protease. The activation mechanism is not known but may involve binding of *recA* protein to damaged or single-stranded DNA. Two proteins are cleaved by this protease activity. Cleavage of the lambda bacteriophage repressor accounts for the induction of phage growth following UV irradiation of lambda lysogens (81, 82). *recA* protein also cleaves the *lexA* gene product, thus derepressing its own synthesis (83). Some mutants of *lexA* (*lexA*$^-$) block induction of *recA* (44). The *lexA* gene product from these mutants is resistant to the *recA* protease (83), so that *recA* remains repressed even under conditions that would induce *recA* in normal cells.

4. ROLE OF *recA* PROTEIN IN UV-INDUCED GENETIC EXCHANGES

The following evidence strongly favors a role for *recA* protein in DNA exchanges associated with postreplication repair of UV-damaged DNA. First, after UV irradiation of *E. coli*, *recA* protein is synthesized at a very high rate (45, 82, 84, 85). Second, *lexA*$^-$ mutants, which do not induce *recA* protein, are deficient in postreplication repair (40). Therefore, high levels of *recA* protein may be required for efficient postreplication repair. And third, *recA* protein binds efficiently and cooperatively to double-stranded DNA containing single-stranded gaps, and binding can then extend into neighboring double-stranded regions through cooperative binding (86). Therefore, *recA* protein might bind to gaps in newly synthesized DNA opposite pyrimidine dimers and create single-stranded ends that could interact with the homologous sister chromosome.

We propose a model, adapted from several previously proposed (18, 39, 87), that describes a role for *recA* protein in the genetic exchanges associated with postreplication repair. As shown in Fig. 2c, *recA* protein binds to the single-stranded region opposite a pyrimidine dimer in a recently replicated DNA molecule from an irradiated cell. Further cooperative binding of recA protein unwinds the double-stranded DNA at the ends of the gap (Fig. 2d). The resulting free single-strands, with recA protein bound to them, then interact with a sister chromosome and search for homologous base sequences. When homology is found, D-loop structures are formed (Fig. 2e). Resolution of the D-loop by nucleolytic removal of the displaced parental strand and ligation of the parental DNA strand to the invading strand leads to the formation of the intermediate structure shown in Fig. 2f. Then DNA synthesis displaces the transferred, newly synthesized strand

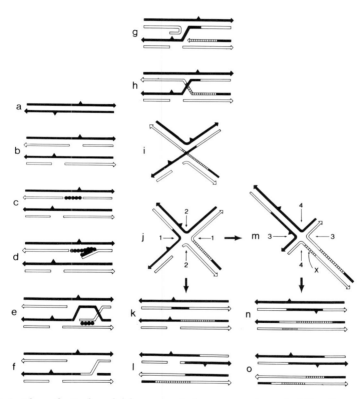

FIG. 2. A hypothetical model for removal of gaps opposite pyrimidine dimers in partially replicated bacterial chromosomes by recA protein-promoted recombination. (a) A bacterial chromosome containing pyrimidine dimers (▲) is depicted. After replication, the two progeny chromosomes contain gaps opposite the dimers (b). The remaining steps are described in the text. Arrows on ends of molecules designate polarity of the DNA strands.

and its covalently attached parental strand, and causes this displaced strand to be assimilated by the other molecule (Fig. 2g). As a result, displacement allows filling in of the gap in the new strand by parental DNA. Additional synthesis could generate displacement of the 5' hydroxyl end of the new DNA at the original gap (Fig. 2g). This displaced DNA could then be ligated to the 3' hydroxyl end of the extended parental strand (Fig. 2h), thus forming a structure originally proposed by Holliday (88) as being an intermediate in recombination.

The Holliday intermediate may be resolved through an isomerization step as first suggested by Ganesan (39). This isomerization is depicted diagrammatically by redrawing the structure shown in Fig. 2h as a cruciform structure (Fig. 2i; see 86). The two lower strands are held fixed while the upper strands are rotated 180 degrees to generate the isomeric molecular structure in Fig. 2j. Cleavage of this structure at position 1 generates the two progeny molecules shown in Fig. 2k, in which the strands outside the region of exchange have returned to the parental, nonrecombinant configuration. Cleavage of the molecule in Fig. 2j at position 2 generates two different progeny molecules in which the flanking parental strands are in a recombinant configuration (Fig. 2l). Figures 2m, 2n, and 2o show a hypothetical sequence for the removal of gaps opposite dimers by branch migration of the structure shown in Fig. 2j. The gap shown on the lower left arm of the structure in Fig. 2j is brought opposite a continuous strand and is filled in, as indicated in Fig. 2m, at the position marked with an X. By this mechanism of branch migration, gaps opposite neighboring pyrimidine dimers might also be removed.

The above model attempts to explain the observations (38, 89) that parental DNA becomes covalently associated with DNA synthesized after irradiation, and that there is approximately one exchange event for every pyrimidine dimer. In addition, isomerization of Holliday intermediates accounts for the association of dimer-containing parental DNA with newly synthesized DNA (39). This model, while it accounts for many of the features of postreplication repair in bacteria, is purely hypothetical and only attempts to deal with structural rearrangements in localized regions of the chromosome. If many such interactions were to take place, complex structural rearrangements of the chromosome might also be necessary. It is also possible that such rearrangements and associated changes in chromosomal replication and cell division are coordinately controlled by the *recA* gene and associated functions (see Section IV,A).

III. DNA Synthesis in Ultraviolet-Irradiated Mammalian Cells

A. Control of DNA Replication

In order to discuss the effects of UV irradiation on DNA synthesis in mammalian cells, it is first necessary to understand the normal mode of replication of mammalian chromosomes and the way this replication differs from that observed in bacteria. An *E. coli* chromosome has a unique origin of replication and replicates its circular chromosome bidirectionally (72). Thus, the entire *E. coli* chromosome, composed of four million base-pairs, is replicated as a single unit, or replicon. By contrast, each chromosome in mammalian cells has multiple origins of replication (90), separated from each other by 60,000 to 180,000 base-pairs (90–92). As illustrated in Fig. 3, replication proceeds bidirectionally from these origins (Fig. 3B) and terminates either at some defined termination sequence or at the position where two replication forks moving in opposite directions happen to meet (Fig. 3C); the precise mechanism of termination has not yet been elucidated. The final step in replication is the joining of progeny molecules synthesized from adjacent origins (Fig. 3D). These individual replication units have been termed "replicons" because they appear to resemble the single unit of control that replicates the *E. coli* chromosome. In summary, replication of mammalian cell chromosomes is organized into many replicon units 200- to 600-fold smaller (depending upon the mammalian cell-type) than the single replicon of the *E. coli* chromosome. It is important to recognize this difference in considering the relative responses of DNA synthesis in mammalian cells and bacteria to DNA damage.

A second difference between chromosome replication in mammalian cells and *E. coli* is that the bacterial chromosome may initiate a new cycle of replication at the origin of replication prior to the completion of a previous round of replication (61). The presence of DNA damage itself can stimulate such a new round of chromosomal replication (Section II,G). In unirradiated mammalian cells, an individual replicon is probably active only once in the cell cycle, based on the following observations. First, clusters of adjacent replicons tend to initiate in synchrony (90), as though under coordinate control (Fig. 3A). Second, an ordered sequence of replicon initiation has been observed, such that clusters of replicons appear to initiate at specific periods during the S phase of the cell cycle (90). Little information is presently available as to whether this normal pattern of DNA replication in mammalian cells can be

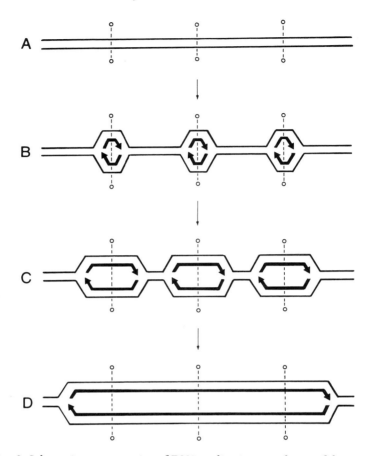

FIG. 3. Schematic representation of DNA replication in a cluster of three replicons in a mammalian chromosome. (A) An unreplicated region of the chromosome contains three origins of replication (designated by o). (B) Replication initiates simultaneously at all three origins and proceeds bidirectionally until (C) newly replicated strands (drawn as heavy lines) from adjacent replicons meet, allowing (D) joining of these strands into DNA lengths greater than that of single replicons. Arrows on newly replicated strands show their polarity.

disrupted by UV damage. However, experiments with synchronized Chinese hamster ovary cells (93) indicate that replicons do not initiate a second time during the same S phase following UV irradiation. Cells were pulse-labeled with tritiated thymidine, exposed to UV, and grown for short periods in unlabeled medium containing dBrU; the DNA was analyzed by density gradient centrifugation. The tritium-labeled DNA, which had replicated prior to UV treatment, was found only at the fully light density, as would be expected for DNA that had not replicated during the period of growth in dBrU. Consequently, there were no detectable second rounds of replication following UV irradiation.

B. Effects on the Cell Cycle

Because mammalian cells synthesize DNA during only a fraction of the cell cycle, it is necessary to consider the effects of DNA-damaging treatment on cells in different parts of the cycle. Cell lines of several species, irradiated in G_1, move into S at a normal rate (reviewed in 94). Consequently, the onset of DNA synthesis in these cells is not inhibited by UV irradiation. Cells irradiated in G_2 either progress normally through the cell cycle, or fail to undergo mitosis. However, the fraction of cells blocked in G_2 is small at doses of 10 J/m^2 or less. Presumably then, except for this small fraction blocked in G_2, most cells in an irradiated culture progress into S phase at a nearly normal rate, provided that the experiment is done at high survival levels. In contrast, cells irradiated in S phase move abnormally slowly through S phase while undergoing an associated slowing of DNA synthesis. Some cells become permanently blocked in S and die. Presumably, irradiated cells that must pass through other parts of the cell cycle before entering S will experience this same extension of S phase. Thus, irradiated, exponentially growing cultures of mammalian cells tend to accumulate cells in the S phase of the cell cycle (94).

C. Inhibition of DNA Synthesis

It has generally been observed that UV irradiation of mammalian cells inhibits DNA synthesis. This inhibition is presumably due to the presence of pyrimidine dimers in the chromosomes of these cells, although other types of DNA damage may also play a role in this inhibition. Because there are a large number of replicons in mammalian cell chromosomes, and because they initiate sequentially during S phase, new initiation events might allow relatively high levels of synthesis in irradiated cultures, even though synthesis eventually stops in many replicons. Therefore, it is important to understand the effects of UV on initiation and continuation of DNA synthesis in mammalian cells. Based on the analysis described above (Section III,B), the observed inhibition of DNA synthesis in mammalian cells is probably not due to a block that prevents cells entering S phase. Moreover, because cells in an irradiated culture tend to accumulate in S phase, the extent of inhibition of DNA synthesis in individual cells may be greater than that suggested by comparison of irradiated and unirradiated cultures.

Inhibition of DNA synthesis by low doses of UV (10 J/m^2 or less) has been reported by several laboratories. For example, the extent of inhibition of DNA synthesis in Chinese hamster cells following UV irradiation depends on both the dose of UV and the time after irradiation that the rate of DNA synthesis is measured (95). At a dose of 5 J/m^2, the rate of incorporation of tritium-labeled thymidine was initially comparable to that of unirradiated cells but then decreased steadily between 0 and 30 minutes after irradiation, reaching a value approximately 45% of the initial one. Between 30 and 60 minutes after irradiation, little further change was observed. In a similar experiment with normal human fibroblasts irradiated with this same UV dose, the extent of thymidine incorporation in cells pulse-labeled immediately after irradiation was slightly less than that observed for unirradiated cells, subsequently decreasing to a minimum of 30% to 40% of that for unirradiated cells at 1–5 hours after irradiation (96, 97). Cultures incubated for longer times gradually recovered the ability to synthesize DNA at a normal rate.

This inhibition of DNA synthesis in mammalian cells by UV irradiation may result from any one of a number of possible perturbations of normal DNA replication. First, the frequency of initiation of DNA synthesis in individual replicons might be reduced. Second, the rate of elongation of nascent DNA strands might be slowed. This effect on elongation might result directly from UV damage, since the damaged chromosome itself may not be an efficient template for DNA synthesis. Alternatively, enzymes utilized in normal, semiconservative replication might be sequestered for DNA repair and thus might not be available in concentrations adequate for normal replication. A third possibility is that mammalian cells may possess mechanisms for coordination of DNA replication with DNA repair, such that replication is inhibited while certain repair processes are active. Fourth, UV might alter DNA replication in mammalian cells through induction of novel modes of DNA replication, as occurs in bacteria (see Section II,G). Because one or more of these mechanisms might result in reduced levels of DNA synthesis in irradiated mammalian cells, UV-induced inhibition of DNA synthesis in these cells must be interpreted with caution. Moreover, the relative influence of these mechanisms on DNA synthesis might vary with time after the irradiation treatment. Finally, care must be taken in too readily generalizing observations made on only one mammalian cell type. The effect of UV on DNA synthesis may, in fact, be different in different mammalian cells owing to genetic differences, such as replicon size or properties of the DNA repair or DNA replication enzymes. In the following discussion, we attempt to distinguish among some of these possibilities.

D. Inhibition of Replicon Initiation

Several investigators have attempted to distinguish the consequences of UV irradiation on replicon initiation from those on DNA elongation. For example, Doniger (95) measured the effects of UV irradiation on these parameters in Chinese hamster cells. He pulse-labeled unirradiated and irradiated cultures with [^{3}H]thymidine at the time (25 minutes after irradiation) when the overall rate of DNA synthesis had reached a level 45% of that in unirradiated cells, he then determined, by fiber autoradiography, the distances between replication forks initiated from the same origin. In these studies, a recently replicated region appeared as a tandem pair of labeled segments, separated by an unlabeled central portion synthesized prior to the addition of label and containing the replication origin. The length of this unlabeled region should be proportional to the time between replicon initiation and the addition of label. Doniger observed that the distribution of lengths of these unlabeled regions are similar for both unirradiated and irradiated cells. More important, the distribution in irradiated cells included the same proportion of short regions as found in unirradiated cells, indicating that new replicons were still being initiated at a time when DNA synthesis was reduced 2.5-fold. These results suggest that the population of replicons undergoing replication after UV irradiation in Chinese hamster cells both initiated and replicated at nearly normal rates.

However, the autoradiographic technique used in this study (95) does not provide an indication as to what fraction of the total number of replicons in irradiated cells are functioning compared to the fraction in untreated cells. Consequently, these experiments might not have detected a reduction in the total number of replicons undergoing replication at any given time. It is also possible that the initiation events detected were not representative of the entire irradiated cell population. The labeled regions could have resulted from replication of chromosomes or parts of chromosomes less exposed to UV. The UV doses in these experiments induced only a few pyrimidine dimers per replicon. The actual number in each replicon should vary according to a Poisson distribution. Consequently, slight variations in the dimer yield in parts of individual chromosomes could result in some replicons being free of dimers, and others being more extensively damaged. Finally, this analysis might be faulty since only those replicons were scored in which both replication forks were actively carrying out DNA synthesis. Replicons in which one or both forks might be stalled were not considered. Hence, this analysis may have favored replicons receiving a minimum of DNA damage.

Park and Cleaver (96) examined the ability of human cells to initiate DNA synthesis after UV irradiation by measuring the amount of DNA synthesized per unit time and the size of the newly synthesized DNA (alkaline sucrose gradients). Both parameters decreased early after irradiation and later increased, eventually reaching values similar to those found for unirradiated cells. It is generally assumed (see Section III,E) that the smaller DNA observed soon after irradiation results from a decreased elongation rate of nascent DNA chains due to either partial or complete blockage of the replication forks at sites of pyrimidine dimers in the parental DNA. The recovery of a nearly normal size distribution at later times has been taken to indicate that DNA strands are being elongated normally at these times, despite the continued presence of pyrimidine dimers. In the above work (96), the ability to synthesize DNA strands of normal size recovered before the ability to synthesize normal amounts of DNA. Cleaver and Park suggested that the continued inhibition of DNA synthesis at this time might arise from fewer than normal numbers of replicons being active at a given time. However, this conclusion is based upon indirect evidence and requires verification by more direct measurements of DNA initiation rates.

A further difficulty in interpreting the above experiments (95–97) is that they do not distinguish between normal initiation and abnormal initiation events in replication. As discussed previously (Section III,A), unirradiated cells initiate DNA synthesis from predetermined origin sequences at designated times during the S phase (90). While irradiated cells might initiate DNA synthesis by this same normal mechanism, they might also utilize abnormal replication mechanisms. For example, they might initiate replication from normal replicon origins, but not at the scheduled time during S phase, or they might initiate at abnormal chromosomal locations.

To summarize the effects of UV irradiation on DNA replication in mammalian cells, the above results (95–97) suggest that irradiation does not have a major inhibitory effect on initiation in hamster cells, but may reduce the frequency of DNA initiation in normal human cells. However, owing to the limitations of the techniques employed, further studies are required before any firm conclusions can be reached.

Studies of the effects of UV irradiation on initiation of DNA synthesis are simplified by using a small virus, such as simian virus 40 (SV40), with a simple DNA replication control mechanism and a single origin (98). Sarasin and Hanawalt (99) found that new rounds of replication on UV-irradiated SV40 are initiated in a broad region that includes the normal origin of replication of this virus. Restriction enzyme analysis indicated that this region is preferentially labeled in both unirradiated and irradiated cells infected with virus. Initiation therefore appears to occur at the normal region, although initiation at new sites near the normal origin cannot be totally ruled out by this experiment. The SV40 genome is small enough to code for only a few proteins (99a–99c), and, consequently, this virus may utilize many host-cell functions to carry out replication and repair of viral DNA. Therefore, studies utilizing this virus may provide information concerning the response of these host functions to DNA damage. However, such information may be limited in that it may not accurately reflect the cellular response to damaged host chromatin.

E. Inhibition of DNA Elongation

The experiments (95–97) analyzed above reveal a reduction in the overall rate of DNA synthesis at early times after UV irradiation in both human and rodent cells. This reduction appears not to arise from a reduced fraction of cells in S phase but may result, in part, from inhibition of replicon initiation. A third mechanism for inhibition of DNA synthesis might be a reduction in rates of chain elongation. Decreased rates of elongation might occur by a general slowing of replication fork movement in damaged cells, if, for example, components of the replication complex were sequestered for DNA repair processes. Alternatively, a reduction in chain elongation might occur by specific blockage of replication forks when they reach pyrimidine dimers or other UV-induced lesions. This blockage could be temporary or might exist for more extended periods.

In order to study the effect of UV irradiation on rates of DNA synthesis, Doniger (95) measured the rate of fork movement in UV-irradiated Chinese hamster cells by fiber autoradiography. Irradiated cells were pulse-labeled, the lengths of the labeled regions were measured, and the means of these lengths were calculated as a function of labeling time. These measurements revealed that the average rate of fork movement is not inhibited by a dose of 5 J/m^2 but is 58% inhibited by a dose of 10 J/m^2.

In other experiments (*100*), the ends of growing DNA chains in cells previously irradiated were pulse-labeled with radioactive dBrU. The length of the DNA synthesized during the pulse was then determined by irradiation with 313 nm light, which breaks dBrU-containing regions of DNA. Measured by sedimentation of the DNA in alkaline sucrose gradients, the rate of breakage was slightly slower for DNA from irradiated cells than for DNA from unirradiated cells, suggesting that the molecules from irradiated cells contained slightly shorter dBrU-labeled regions. For UV doses up to 11 J/m^2, the results are consistent with a small reduction in chain elongation in irradiated cells.

In summary, these results demonstrate that a small inhibition of chain elongation is associated with UV damage. It is not possible to determine from these experiments whether these reductions in chain growth occur by a general slowing of the fork movement along all regions of the DNA, or whether fork movement is specifically blocked at sites of UV damage in the parental DNA strands. Moreover, the small reduction in chain elongation rates observed is not sufficient to account for the substantial overall decrease in DNA synthesis in these same irradiated cell cultures (95–97). Therefore, UV must have inhibitory effects of DNA synthesis not detected by these techniques. To investigate further the mechanism of inhibition of DNA synthesis by UV damage, the structures of the replication intermediates in UV-irradiated cells have been analyzed, as described below.

F. Evidence for Blockage of DNA Synthesis at Pyrimidine Dimers

Many investigators have proposed that DNA synthesis is inhibited at sites of pyrimidine dimers in mammalian cells, but that DNA replication resumes downstream from a pyrimidine dimer leaving a gap opposite the dimer, as in bacteria (Section II,C). However, a second possibility (*22, 101*) is that dimers block replication for extended periods without reinitiation. Such extreme inhibition would presumably result in progressively fewer replicons being replicated following irradiation and could account for the observed decrease in DNA synthesis (95, 96). In this and in Sections III,G–J, we review the evidence in favor of this latter viewpoint.

1. ALKALINE SUCROSE GRADIENT ANALYSIS

The experimental technique most commonly employed for studying effects of UV on DNA synthesis in mammalian cells is gradient centrifugation in alkaline sucrose of DNA extracted from cells pulse-labeled shortly after UV irradiation. The molecular weight (size) distribution of the newly synthesized DNA is thus compared to that of unirradiated cells. The results reveal that the DNA synthesized in UV-irradiated mammalian cells has a lower average molecular weight (smaller size) than that in unirradiated cells (e.g., *102–104*). This result is qualitatively similar to that observed in bacteria.

In order to interpret these size determinations, we must first understand those obtained by the same analysis on unirradiated cells. Varying the length of the pulse of radioactive thymidine given to an unirradiated cell culture reveals that DNA synthesized during shorter pulses (15 minutes or less) is smaller than DNA synthesized during longer pulses (30 minutes or more) (*102, 105*). In addition, if a short pulse is followed by a period of growth without label, the label initially in smaller DNA strands appears later in a larger form, similar to that found with longer pulses (*102, 105*). Therefore, DNA strands of larger size synthesized in unirradiated cells are derived from smaller precursors.

The process of elongation of the smaller molecules detected in alkaline sucrose gradients probably represents the joining of progeny DNA strands synthesized in two adjacent replicons into longer DNA molecules (*102*). As indicated in Fig. 2, DNA synthesis is believed to initiate simultaneously at the origins of a tandem set of replicons, and to proceed bidirectionally until sufficient synthesis has occurred to allow joining of DNA molecules from adjacent replicons (*90*). Adjacent replicon origins are approximately 20–60 μm apart, depending on the type of mammalian cell (*90–92, 95*), and it takes approximately 30–90 minutes to complete replication of a single replicon (*95*). As a result, molecules labeled during a still longer pulse should have a size distribution that varies from that of short strands associated with new initiations at the end of the labeling period to that of long strands resulting from joining of progeny strands from two or more adjacent replicons. If the labeling period is shorter than the time needed to replicate the average replicon, the size distribution should shift to smaller values, as illustrated in Fig. 2.

A possible explanation for the observed smaller size of the DNA synthesized in irradiated cultures is that individual replicons in these cells are not as fully replicated as they would be in unirradiated cells pulse-labeled for a comparable period, due to a decreased rate of chain elongation. Evidence against this possibility comes from experiments in which the pulse-labeling period for irradiated cells is increased until the level of incorporation is the same as that in the unirradiated cell population. Under these conditions, the average size of the newly synthesized DNA in irradiated cells shifts upward, but still remains less than that in unirradiated cells (95, 103, 105). Therefore, variations in chain elongation rates cannot account totally for the smaller size of DNA synthesized in irradiated cells.

It will be recalled that, in bacteria, the average size of the newly synthesized DNA in experiments similar to those described above is roughly proportional to the average distance between adjacent pyrimidine dimers on the same DNA strand; also these short DNA strands are as abundant whether the labeling pulse comes immediately after irradiation, or 10–40 minutes later (Section II,B). However, in mammalian cells, DNA labeled immediately after irradiation contains more of the larger species than DNA labeled in a pulse given 20 minutes (or longer) after irradiation (102). A reasonable explanation is that unlabeled strands synthesized prior to irradiation may be elongated during the pulse following irradiation. These end-labeled molecules would then have a size distribution reflecting the structure of DNA strands synthesized in part on an undamaged DNA template, and in part on a damaged template. However, if pulse-labeling of irradiated cells is delayed for 20–60 minutes (the time required for synthesis of an average mammalian replicon) after UV exposure, the majority of nascent DNA strands made prior to irradiation should be elongated up to natural termination sequences or to UV-induced replication blocks, and should not become labeled. The size of the DNA made after this period should be more representative of replication on UV-damaged DNA templates.

Using this labeling protocol, investigators have attempted to determine whether pyrimidine dimers block chain elongation in UV-irradiated mammalian cells by comparing the sizes of DNA synthesized in these cells with the estimated average distance between pyrimidine dimers. A correspondence between these two parameters would suggest that DNA synthesis is halted at the sites of pyrimidine dimers. For a given mammalian cell type, such correspondences have been reported by some (95, 102–104), but others have observed that the average size of the newly synthesized strands exceeds that of the interdimer distance (106, 107).

The results from these studies vary with the length of the pulse of radioactive label given the irradiated cells and with the type of mammalian cell studied. Shorter pulse lengths tend to yield smaller DNAs and, in some cases, DNAs of sizes comparable to the average interdimer distances (95, 102, 103). Since replicon size differs in different cell types, cells with longer replicons should contain, on the average, more dimers in each replicon for a given dose of UV. Consequently, the size of the DNA synthesized should be more drastically affected in these cells than in cells with more closely spaced replicons. Cleaver (4) has suggested that the greater difficulty encountered by some investigators (56) in detecting size differences between nascent DNA from UV-irradiated human cells and those from untreated cells could be due to the short interreplicon distance in these cells. The genetic background of a given cell type also affects the results obtained. For example, DNA synthesized shortly after UV irradiation in wild-type human cells appears to be significantly larger than DNA from irradiated XP variant cells (56, 57, 107).

In addition to the variations described above, alkaline sucrose gradient analyses of the size of nascent DNA are subject to certain technical limitations, such as: (a) errors in estimations of the average M_r's of DNA from peaks (which are often broad); (b) errors in calculations of the M_r's of smaller molecules, which sediment, under the conditions used, in the upper portion of the gradient, where the position of a DNA molecule varies rapidly with M_r; and (c) errors in the estimation of the average interdimer distances, since this estimation assumes that pyrimidine dimers are induced at equal frequencies in all parts of the chromosomes. If, in fact, some regions of the chromosome have fewer dimers than other regions, the DNA synthesis detected in these experiments might be occurring preferentially in these less-damaged regions.

Consideration of these difficulties makes it impossible for this technique to show with certainty that pyrimidine dimers block replication in all UV-irradiated mammalian cells. However, the weight of the evidence supports the conclusion that, as in bacteria, mammalian replication forks are blocked at all or many sites of UV damage in the template strands.

2. FIBER AUTORADIOGRAPHY

Studies of DNA replication in UV-irradiated human (HeLa) cells by fiber autoradiography (101) also indicate that DNA synthesis may be blocked at sites of pyrimidine dimers. When cells were pulse-labeled immediately after UV irradiation for times up to 90 minutes, and the distributions of the lengths of the DNA synthesized were determined, it appeared that replication proceeds normally for approximately 20 minutes after irradiation, and that longer pulses gave no further increases in the average lengths of DNA synthesized. Furthermore, the average length of the DNA made on irradiated templates was approximately equal to the average distance between pyrimidine dimers. It was concluded from these studies that replication in irradiated HeLa cells continues until a dimer is reached, and then remains blocked for periods of up to 90 minutes.

These studies contrast with fiber autoradiograph analyses of irradiated DNA from Chinese hamster cells (95). In these cells, the average length of the DNA synthesized between 25 and 85 minutes after UV irradiation appears equivalent to the distance between several pyrimidine dimers. On the other hand, alkaline sucrose gradient analysis of the DNA synthesized in hamster cells during short pulses (15 minutes) showed that this DNA is approximately equal to the size expected from the average distance between pyrimidine dimers. Taken together, these results suggest that, in Chinese hamster cells, DNA synthesis may be blocked at sites of pyrimidine dimers for brief periods, but may rapidly resume, either by continuation of replication past the damaged sites or by discontinuous synthesis, which leaves a gap opposite the pyrimidine dimer. (Such gaps would not be detected by the fiber autoradiography analysis.) Consequently, a comparison of these results (95) and those for HeLa cells (101) suggests that human and hamster cells may differ with regard to the severity of inhibition of DNA synthesis by DNA damage in the template strands. However, the criterion used in the study of Chinese hamster cells (95) for selecting fibers to be measured was tandem sets of symmetrical pairs. Since this criterion excludes replicons in which one or both replication forks are blocked at sites of UV damage, the difference between hamster and human cells may not be as great as suggested by the data. In addition, the human cells (101) were examined at times after irradiation when DNA synthesis was still decreasing, while the hamster cells (95) were examined when DNA synthesis had decreased to a constant level. Therefore, the differences in these studies might be due to differences in the times after UV at which the cells were examined. It is possible that replication forks or chromatin structure could be gradually modified following DNA damage in such a way as to allow more efficient replication of damaged DNA. Such changes might already have occurred in the hamster cell study (95), but not in the study of HeLa cells (101).

3. STUDIES OF VIRAL REPLICATION

Possibly the most definitive experiments suggesting that DNA synthesis can be blocked at sites of pyrimidine dimers in mammalian cells are those of Sarasin and Hanawalt (99), who examined the effects of UV irradiation on viral DNA replication in SV40-infected monkey cells. To synchronize viral replication, they employed a temperature-sensitive mutant of SV40 (*tsA58*) that is blocked in initiation of DNA replication at the nonpermissive growth temperature. Incubating infected cells at high temperature allows accumulation of fully replicated viral genomes. Subsequent irradiation of these cells followed by incubation at the permissive temperature allowed a study of replication of a synchronized population of molecules. The size of the new DNA in this population, analyzed in alkaline sucrose gradients, was approximately equal to the average distance between two dimers on a parental DNA template. This result was consistent with replication of a SV40 molecule in which both newly synthesized strands are replicated bidirectionally from the origin until a pyrimidine dimer is encountered in the parental strand at each end of the nascent strands. A similar conclusion was reached by Williams and Cleaver (108). These experiments, while subject to errors in estimation of size and average interdimer distances, are simpler to interpret than experiments with uninfected cells, because they employ the small, well defined template of the SV40 virus, which contains a single replication origin and undergoes replication by a relatively well understood mechanism.

There is additional evidence that SV40 replication may be blocked at the sites of DNA damage. Cells were infected with mutant *tsA58* and allowed to accumulate nonreplicating viral chromosomes by incubation at the nonpermissive temperature (99). After UV irradiation, infected cultures were pulse-labeled at the permissive temperature, and the viral DNA was extracted and subjected to restriction enzyme analysis, which revealed that in both UV-irradiated and unirradiated cells, label was preferentially incorporated into DNA sequences that included the origin of replication. DNA sequences further removed from this origin contained proportionately less label. In the unirradiated virus, the ratio of the label incorporated near the origin to that incorporated in DNA sequences further removed from the origin was greater than one, indicating that some of these molecules were not fully replicated. For UV-damaged virus, this ratio was higher than that for undamaged virus, suggesting that replication of irradiated SV40 initiates at or near the normal origin of replication and proceeds bidirectionally until blocked, presumably by UV damage in the template strands. Consequently, fewer UV-irradiated viral molecules complete the replication cycle than do virus genomes from unirradiated cultures.

4. MODEL FOR INHIBITION OF THE REPLICATION COMPLEX AT DAMAGED SITES

As judged by the sizes of DNA molecules synthesized shortly after irradiation, mammalian cells appear unable to synthesize continuous

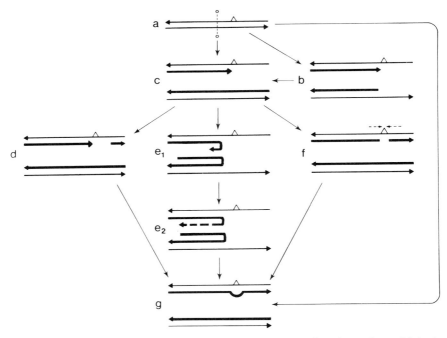

FIG. 4. Proposed modes of DNA replication on UV-irradiated templates. (a) A single, unreplicated replicon is depicted with origin (o) and a pyrimidine dimer (△). Replication is initiated at the origin and proceeds bidirectionally according to schemes b–g. (b) Replication may be blocked in both nascent strands by a dimer in only one parental strand, or (c) replication of only one strand may be blocked by the dimer, while synthesis of the sister strand may continue unimpeded. (d) Synthesis of the nascent strand, blocked at the dimer, may continue by reinitiation beyond the dimer, leaving a gap. Alternatively, replication might proceed continuously past the pyrimidine dimer by a copy-choice, branch-migration model (e_1 and e_2), in which the nascent strands pair and the blocked strand uses the other nascent strand as a template. Finally, the pyrimidine dimer may block replication for extended periods, allowing a nascent strand from the adjacent replicon to synthesize up to the dimer (from the right) (f). The dashed arrows in (f) indicate the convergence of synthesis of the two strands blocked at the pyrimidine dimer. The structures in (a), (d), (e_2), and (f) may eventually be resolved as shown in (g) by mechanisms discussed in the text. Arrows on ends of molecules indicate strand polarity.

DNA strands past sites of pyrimidine dimers. DNA synthesis might be blocked at dimer sites in one of two ways, as illustrated in Fig. 4 (b and c). First, replication might be inhibited in both nascent strands of a given replication fork (Fig. 4b), even though damage appears in the parental strand opposite only one of these nascent strands. Alternatively, replication might be blocked only in the nascent strand complementary to the parental strand containing the pyrimidine dimer, while the other nascent strand on the sister molecule might continue to replicate unimpeded (Fig. 4c). One or both of these mechanisms could apply to the same cell, depending upon the direction of synthesis and the polarity of the template strand. Evidence in favor of this second mode comes from studies on DNA synthesis of UV-irradiated SV40 virus (99) in which, as described above, the size of the viral DNA synthesized during a short labeling pulse after UV irradiation had an average length comparable to the estimated average length between pyrimidine dimers on one strand of viral DNA. If both nascent strands in a given replication fork were blocked at a pyrimidine dimer present in only one of the parental strands, the expected length of the nascent DNA should be approximately half of the interdimer distance. These results, therefore, favor a mechanism whereby DNA replication of a given molecule is blocked only by the presence of DNA damage in the template. However, since the experimental error in the estimates of interdimer distances is large, alternative explanations for these results must also be considered.

G. Replication Past Sites of UV Damage

While replication in mammalian cells appears to be blocked initially at sites of UV-induced damage, there is substantial evidence that these cells eventually are able to replicate past these lesions. First, the entire chromosome complement of Chinese hamster ovary cells replicates within 24 hours after exposure to low doses of UV, as judged by the extent of density change following a period of growth in density-labeled DNA precursors (93). Since little, if any, excision of pyrimidine dimers take place in these cells, they appear able to complete replication of chromosomes containing pyrimidine dimers. In addition, other cells unable to excise pyrimidine dimers (mouse L5178Y cells or human XP cells) undergo many cell divisions following UV irradiation, despite the continued presence of UV damage (94, 102).

As discussed in Section III,E, Doniger (95) measured the average length of DNA synthesized in irradiated cells during a given labeling time, and determined the average yield of pyrimidine dimers in irradiated chromosomes at each dose tested in order to estimate how frequently replication forks might pass pyrimidine dimers. This analysis demonstrated that for labeling periods of 60 minutes, the length of DNA replicated must have passed several pyrimidine dimers, assuming that the dimer frequency was the same in all replicated regions of the chromosome. However (see Section III,E), the technique used may have preferentially analyzed regions of the chromosome containing few pyrimidine dimers.

Analyses by alkaline sucrose sedimentation also suggest that DNA replication past damaged sites may occur (*102, 103, 105, 109, 110*). As discussed in Section III,F, DNA pulse-labeled shortly after irradiation of rodent or human cells is smaller than DNA synthesized in unirradiated cells, presumably because of blockage at sites of pyrimidine dimers. If these same irradiated cells are incubated further in unlabeled medium before analysis, the label that appeared initially in smaller molecules shifts to larger ones, suggesting that the cells are eventually able to overcome the replication block.

These experiments suggest that DNA replication can proceed past pyrimidine dimers in some mammalian cells. Several possible mechanisms whereby such replication might occur are illustrated in Fig. 4. One is that proposed for bacteria (Fig. 4d), in which replication of UV-damaged templates is blocked at sites of pyrimidine dimers, but is reinitiated downstream from the dimers, leaving gaps in the new strands opposite the dimers. The fact that DNA in irradiated mammalian cells is initially synthesized as smaller molecules that are subsequently elongated (or joined) to form normal-sized molecules has led many investigators to assume that the bacterial model for postreplication repair is also correct for mammalian cells.

However, there is quite convincing evidence that this prokaryotic model may not be applicable to mammalian cells. It might be argued that the relatively small inhibitory effect of UV on incorporation of labeled precursor into most mammalian cells is evidence that blockage of replication forks at pyrimidine dimers is temporary, and that reinitiation of DNA synthesis occurs at sites past the dimers. However, the large number of replicons in mammalian cells and their sequential initiation during S phase could also maintain high levels of DNA synthesis in irradiated cells even if dimers were absolute blocks to replication in these cells. Analysis of the structure of DNA synthesized in irradiated human and rodent cells (described in the next two sections) suggests, although rather indirectly, that replication forks are blocked at pyrimidine dimers, and that reinitiation downstream may not occur.

H. Do Gaps Opposite Pyrimidine Dimers Exist in Replicated DNA?

If DNA synthesis in mammalian cells leaves gaps opposite pyrimidine dimers, it should be possible to detect the gaps in the nascent strand opposite these lesions. Experiments to detect such structures have been performed with mouse L5178Y cells (*102*) and with human fibroblasts (*111*). The cells were pulse-labeled with radioactive thymidine soon after UV irradiation and subsequently incubated in nonradioactive medium containing dBrU. In alkaline sucrose gradients, the DNA isolated from cells immediately after the pulse period appeared reproducibly smaller than the DNA from unirradiated cells (see Section III,G). Molecules from cells first pulse-labeled and then incubated in unlabeled medium containing dBrU sedimented at a position of nearly normal size, indicating that the short DNA molecules synthesized after UV irradiation had been joined or elongated into longer molecules (Section III,G). However, when this longer DNA was exposed to 313 nm light to break the DNA in regions containing dBrU, the profile shifted to a lower size distribution similar to that obtained at earlier times. These results are consistent with two possible mechanisms for replication of DNA on UV-irradiated templates: (a) continuous replication past pyrimidine dimers, using the smaller molecules as primers; or (b) discontinuous replication and later filling of the gaps by *de novo* DNA synthesis. These possibilities were distinguished by examining the dose of 313 nm light required to break regions of DNA containing dBrU and by calculating from these data the length of the dBrU-containing DNA. This calculation revealed lengths of approximately 400–1600 bases. Therefore, since the incubation period in the presence of dBrU was sufficient to allow DNA replication of entire replicons (60,000–180,000 bases), it was concluded that these short regions of dBrU-containing DNA in irradiated cells originated from the filling of gaps by *de novo* synthesis.

An objection (22) to this conclusion is that the thymine precursor pools in UV-irradiated cells may be altered, so that dBrU substitution in DNA synthesized following exposure to UV light occurs at abnormally low levels. Consequently, the lengths of DNA synthesized under these conditions might actually be much longer than estimated, and might have resulted from end-addition onto pulse-labeled molecules rather than gap-filling. Since further experiments have not been performed to investigate this objection, it is difficult to draw a firm conclusion about the photolysis experiment described above.

A different approach to detecting gaps in newly replicated mammalian DNA is the measurement of the susceptibility of newly replicated DNA molecules from UV-irradiated human fibroblasts (strain WI38) to S1 endonuclease (specific for single-stranded DNA) from *Neurospora crassa* (112). From the finding that these molecules containing UV damage are abnormally sensitive to S1, it was concluded (112) that this DNA contains single-stranded regions, but it was not determined where these gaps were positioned relative to the pyrimidine dimers. In other experiments (104, 113), the new DNA from irradiated WI38 cells was treated with the UV-specific endonuclease (T4 endonuclease V), which makes single-strand incisions next to pyrimidine dimers in both single-stranded and double-stranded DNA (6). It was anticipated that this enzyme should also cut DNA with single-stranded regions opposite dimers, thereby effectively making double-stranded breaks in the molecule. However, no double-stranded breaks were detected, indicating a possible absence of gaps opposite dimers (*104, 113*). A difficulty with this argument is that the activity of the T4 enzyme toward the presumed DNA structure is unknown. It has not been shown, for example, that this method can demonstrate dimer-gap structures in bacterial systems, where the existence of such structures is strongly suggested by independent methods. Consequently, the experiments in human cells with the T4 endonuclease V are inconclusive at present. However, if the T4 enzyme can be shown to have the required specificity, the above results would imply that partially replicated molecules with single-stranded regions accumulate in UV-irradiated human cells, but that the single-stranded regions are not opposite pyrimidine dimers.

I. Does Genetic Recombination Play a Role in DNA Repair or in Mechanisms of Tolerance of Ultraviolet Damage?

Returning to the bacterial model for postreplication repair, the gaps left opposite dimers during replication of irradiated DNA are believed to be filled in by recombination with undamaged homologous DNA from sister chromosomes. Evidence that gap-filling occurs by a recombination mechanism has been obtained in bacteria using three experimental approaches (Section II,D). First, transfer of pyrimidine dimers from parental DNA into newly synthesized DNA molecules has been detected (39), probably resulting from DNA exchanges between parental and nascent DNA strands. Second, UV-induced DNA exchanges have been demonstrated by means of density-labeled DNA. And third, UV-induced genetic exchanges during genetic crosses between bacteriophages or *E. coli* cells have been observed. Similar experiments have been carried out to detect UV-induced recombination in mammalian cells.

Experiments that attempt to demonstrate the transfer of pyrimidine dimers from parental to newly synthesized strands in UV-irradiated human fibroblasts have been performed, using the sensitivity of nascent DNA molecules to UV-specific endonucleases as an assay for the presence of dimers (*113–116*). The results suggest that this transfer may occur at a low level (5 to 15% of the total pyrimidine dimers), whereas in *E. coli* 50% of the dimers are transferred (39).

However, the appearance of pyrimidine dimers in newly synthesized DNA of human cells may arise not from recombination, but rather from end-addition of new DNA chains onto molecules synthesized prior to irradiation (*116*). As discussed earlier (Section III,F), pulse-labeling of DNA shortly after irradiation yields two classes of labeled molecules: (*a*) short strands initiated after the UV treatment, labeled along their entire length; and (*b*) longer strands initiated prior to irradiation, labeled only at the ends. A delay between the times of irradiation and pulse-labeling, for a period approximately equal to the time required to complete synthesis of replicons, should yield labeled molecules initiated primarily after UV irradiation. If irradiated cells are incubated in unlabeled medium for longer and longer periods prior to pulse-labeling, the fraction of pulse-labeled molecules susceptible to cleavage by the UV specific endonucleases progressively decreases (*116*). This suggests that transfer of pyrimidine dimers to newly synthesized DNA strands does not occur in human cells. However, other investigators (*115*) have detected pyrimidine dimers associated with newly synthesized DNA in UV-irradiated human cells not labeled until 2 hours after UV treatment. Therefore, the possibility that transfer of pyrimidine dimers to nascent DNA strands may occur at a low level in human cells cannot presently be eliminated.

The observation that extensive transfer of dimers to nascent DNA strands does not occur in mammalian cells at the high level detected in bacteria also does not necessarily eliminate a recombination mechanism for gap-filling. As shown in Fig. 2, modes of genetic recombination in which dimer transfer does not occur are possible. Very short regions compared to the average distance between pyrimidine dimers, might be exchanged, thus reducing the chance of dimer transfer to newly synthesized DNA.

Experiments to detect UV-induced DNA exchanges with density-labeled DNA have also been inconclusive owing to difficulties in interpreting the DNA structures detected by this analysis (see Section III,J). In one set of experiments (*117*), Chinese hamster cells were synchronized in G_1, grown in medium containing unlabeled dBrU for 2.5 hours, UV-irradiated, and transferred to medium containing tritium-labeled dBrU for 4 hours. The total incubation period (6.5 hours) in media containing dBrU was sufficiently shorter than the time between the end of one S phase and the beginning of the next (7 hours) that a second round of replication could not have occurred. Consequently, labeled DNA containing dBrU in both strands could not have been produced by reinitiation of DNA replication. The DNA was analyzed in neutral cesium chloride density gradients for the presence of tritium-labeled DNA exhibiting the density of fully heavy DNA. A small number of such fully heavy molecules and also molecules containing both hybrid and fully heavy regions were found in unirradiated cultures, and were observed to increase in frequency fourfold after a UV dose of 10 J/m^2. Control experiments demonstrated that such molecules were not rapidly renaturing palindromic sequences replicated during incubation in the medium containing labeled dBrU. Consequently, these fully heavy molecules and molecules containing short regions of fully heavy density may have resulted from genetic exchanges in which a segment of a light DNA strand in a hybrid molecule was replaced by the homologous segment from a strand containing dBrU. Furthermore, the fact that the formation of this DNA was stimulated by UV suggests that UV damage may induce a low level of DNA exchanges in cellular DNA of hamster cells.

However, other experiments with human cells (*118*) suggest that the formation of such fully heavy DNA following transfer from medium containing light isotopes to medium containing heavy isotopes may not arise from recombination, but rather from a replication intermediate in which pairing occurs between both nascent strands of a given replication fork (Section III,J). The appearance of fully heavy DNA molecules from these cells was transitory, and the number of these molecules decreased following a chase in unlabeled medium. Since the dBrU experiments with Chinese hamster cells (*117*) examined the fully heavy DNA at only one time after UV irradiation, it is not possible to determine whether the appearance of the heavy DNA might also be transitory, as would be expected for a replication intermediate.

Further evidence against a role for recombination in postreplication repair of mammalian cells comes from the experiments (*102*) described previously (Section II,H), in which short, newly replicated DNA strands made in mouse cells after UV irradiation were allowed to elongate in the presence of dBrU. These elongated strands were found to contain regions of dBrU-substituted DNA in lengths that appeared to be consistent with a mechanism involving filling of gaps in nascent DNA with dBrU-containing DNA. We have previously (Section III,H) indicated difficulties in interpreting these experiments. However, if the mechanism proposed is correct, the filling of gaps in the nascent strands of UV irradiated cells arises primarily from *de novo* synthesis, not from recombination.

Although a role for recombination in the repair of UV-induced damage in mammalian cells has been questioned, UV-stimulated genetic recombination has been observed in virus-infected mammalian cells. The formation of recombinant animal viruses increases after irradiation and infection of mammalian cells with pairs of mutant herpes simplex or SV40 viruses (119–121) (Section V,C). In addition, multiplicity reactivation of UV-irradiated herpes simplex, human adeno, and SV40 viruses has been detected in human and monkey cells (120, 122, 123) (see Section V,C). In these experiments, the fraction of infected cells able to produce viable virus following infection at a high multiplicity of infection with damaged virus was substantially higher than that of cells infected at a low multiplicity of infection. Therefore, it appears that interactions between two or more UV-damaged viral genomes in the same cell can promote viral repair. Since UV-stimulated genetic exchanges for these same viral systems have also been reported, it appears likely that multiplicity reactivation may occur by a recombination mechanism, as it does in phage-infected bacteria. However, other mechanisms, such as complementation between damaged viral genomes, cannot be eliminated at present. The possibility that viral DNA might be repaired by a recombination mechanism, suggests that damaged host-cell DNA might also be repaired by recombination. While adeno virus and herpes simplex virus have sufficient genomic capacity to code for viral recombination functions (124, 125), only four gene products of SV40 are known (99a–99c). Consequently, SV40 probably utilizes host-cell functions in order to undergo UV-stimulated recombination or multiplicity reactivation, and these same host functions might also act on UV-damaged cellular DNA.

In summary, studies of the structure of the newly synthesized DNA in irradiated mammalian cells favor temporary blockage of DNA replication forks at sites of pyrimidine dimers. Owing to the diversity of results obtained in different laboratories and on different cell types, it is difficult to draw any firm conclusions as to what mammalian cells typically do next. Hamster and mouse cells might reinitiate synthesis downstream from the dimer, but human cells might not reinitiate DNA synthesis in this manner. Convincing evidence either confirming or disproving the existence of this mechanism is presently lacking. Few, if any, DNA exchanges, which appear to be involved in the filling of gaps in *E. coli*, have been detected in those mammalian cells analyzed for such a mechanism, possibly because such gapped structures are not formed in these cell types. On the other hand, studies with animal cells infected with UV-irradiated viruses suggest that genetic recombination may be important for repair of UV-damaged viral DNA. However, the mechanism by which such recombination might aid in repair or tolerance of UV damage is as yet unknown.

J. Mechanisms for Replication of Damaged DNA

Because several of the features of DNA synthesis in UV-irradiated mammalian cells are very different from those of bacteria, alternative mechanisms for replication of damaged mammalian chromosomes have been proposed. One such model (22, 99, 101) proposes that replication forks, while initially blocked at pyrimidine dimers, resume replication past the dimer at a later time without leaving a gap. Evidence consistent with this idea comes from the observation (99) that viral DNA synthesis in SV40-infected monkey cells initially appeared to be blocked at sites of pyrimidine dimers (see Section III,F), but later resumed without gap formation. The sizes of viral DNA synthesized in pulse-labeled, infected cells, assayed in alkaline sucrose gradients, were approximately equal to the distance between two adjacent dimers on a parental DNA strand. Since the ratio of label incorporated into DNA sequences near the origin of replication over that for sequences farther removed from this origin was higher for irradiated than for unirradiated virus, DNA synthesis did not appear to reinitiate beyond dimers during this time (see Section III,F). If pulse-labeled cultures were subsequently incubated in medium containing unlabeled DNA precursors, pulse-labeled viral molecules, which initially were small, became progressively longer, as though replication past the sites of dimers were eventually occurring in a continuous fashion. Since SV40 genomes undergo frequent interactions with host chromatin (127), it is possible that recombination with host sequences or with other viral genomes might account for the apparent continuation of viral DNA synthesis past damaged regions. However, extensive recombination did not occur in these experiments (99) as dimers were not exchanged between parental and progeny DNA. Consequently, the data (99) are consistent with a mechanism for continuous replication past sites of UV damage in mammalian cells.

Since SV40 presumably uses many host-cell functions for its replication, it appears probable that these same functions are used for host DNA replication. Consequently, the mechanism for replication of UV-damaged viral and host DNA might be the same. However, further experiments are required to establish this possibility more firmly.

A similar elongation of abnormally short DNA molecules synthesized on cellular DNA from irradiated cells, which appears to be due to *de novo* DNA synthesis, has been observed in many mammalian cells (see Section III,G) (102). However, because of the complexities of the cellular systems studied, it is not presently possible to determine whether this elongation results from continuous DNA synthesis, or from discontinuous synthesis in which gaps are later filled by *de novo* DNA synthesis.

Two mechanisms whereby DNA synthesis might replicate continuously past sites of pyrimidine dimers are presented in Fig. 4: (*a*) "copy-choice" involving strand displacement and branch migration (Fig. 4e$_1$ and e$_2$); (*b*) direct insertion of nucleotides opposite dimers (Fig. 4g). In the copy-choice model, originally proposed by Fujiwara and Tatsumi (*118*) and Higgins *et al.* (*128*), replication is blocked in one nascent strand of a given replication fork at the site of a pyrimidine dimer in the parental strand, but replication of the other nascent strand continues. Subsequently, the nascent strands are partially displaced and the parental strands reassociate by a process of branch migration. Concurrently, the nascent strands pair, allowing the strand initially blocked at the pyrimidine dimer to copy the required DNA sequence from the complementary nascent strand. After this short period of DNA synthesis, branch migration again allows the nascent strands to pair with the complementary parental strands, and the normal mechanism of semiconservative DNA synthesis resumes until the next dimer is reached.

Evidence in favor of this model comes from an examination of the DNA from cultures of human fibroblasts incubated after UV irradiation in radioactive medium containing dBrU (*118*). Analysis of the DNA from these cells in neutral CsCl density gradients showed a small amount of fully heavy DNA, which increased after UV irradiation. However, if these same pulse-labeled cultures were incubated further in unlabeled dBrU for 4 hours, the fully heavy DNA, observed after the pulse, disappeared. These results are consistent with the possibility that the fully heavy DNA resulted from a replication intermediate, such as the pairing of nascent DNA strands proposed in the "copy-choice" mechanism.

However, other experiments (*129, 130*) suggest that this fully heavy DNA is actually an artifact of the DNA isolation procedure, since it is not observed when cellular DNA is cross-linked *in vivo* prior to extraction. Furthermore, although this fully heavy DNA does appear to result from newly replicated DNA, it may result from intermediates at sites of initiation of individual replicons rather than at replication forks (*130*). Consequently, although copy-choice replication is attractive as a mechanism for synthesis of DNA in irradiated cells, there is at present no direct proof of its existence. In addition, there is presently no direct evidence for a mechanism that inserts bases opposite pyrimidine dimers in mammalian cells.

An alternative mechanism for continuous replication on mammalian cell chromosomes following UV irradiation is also diagrammed in Fig. 4 (*107*). In this model, replication is blocked at sites of at least some pyrimidine dimers but continues unimpeded along stretches of DNA without dimers. It is not presently known whether synthesis of a strand normally stops when a natural termination sequence is reached, or whether synthesis may continue until a replication fork from the adjacent replicon is encountered (Section III,A). If this second possibility is correct, replication in UV-irradiated cells might continue into neighboring replicons along undamaged stretches of parental DNA. Replication might then be halted only by encountering a pyrimidine dimer that had previously blocked the extension of the nascent strand in this second replicon. If only a few dimers were present in each replicon (as occurs at low UV doses), this mechanism would explain the general observation that DNA synthesized in UV-irradiated cells is initially small, but becomes more normal in size during further incubation (*102, 103, 105, 107, 109, 110*). While explaining how chain elongation occurs in irradiated cells, this model (*107*) does not explain how replication continues past the sites of pyrimidine dimers to produce uninterrupted nascent DNA molecules. Presumably, a recombinational or direct insertion mechanism is required to fill in bases opposite dimers.

In summary, then, the weight of the evidence presently available suggests that DNA synthesis in UV-irradiated mammalian cells, while initially blocked at sites of UV damage, eventually continues past these UV-induced lesions. Whether replication resumes by a continuous or discontinuous mechanism has not been firmly established. However, there is substantial evidence against a discontinuous mechanism and in favor of a continuous one. Although several models for continuous mechanisms have been proposed, none has been verified experimentally. We anticipate that future experiments using small animal viruses as probes, or *in vitro* replication systems derived from mammalian cell components, may be very useful in elucidating the molecular mechanisms of replication of damaged mammalian chromatin.

K. Evidence for and against the Existence of an Inducible Mechanism for Replication of Damaged DNA

Observations with Chinese hamster cells (*131–133*) suggest that replication of UV-damaged, cellular DNA might be facilitated by an inducible mechanism. Cells were exposed to a small dose of UV light several hours prior to a larger dose, and subsequently pulse-labeled and analyzed in alkaline sucrose gradients. The average size of the DNA produced after exposure to the larger dose of UV was less than that observed for DNA from unirradiated cells. This same labeled

DNA increased in size when the cells were subsequently incubated in unlabeled medium. The rate of this increase was slightly greater in cells that had received both the small and the subsequent large dose of UV irradiation than in cells irradiated with only the large dose. This enhanced rate of elongation was not found in cells incubated with cycloheximide between the two UV treatments, indicating a need for *de novo* protein synthesis. These experiments suggest that replication past sites of UV damage may be enhanced by an inducible mechanism. Exposure of the cells to the small UV dose prior to the larger dose might have produced the inducing signal.

The interpretation of the above experiments has been criticized (134) because it fails to take into account the fact that DNA synthesis is drastically reduced in UV-irradiated cells, and is further reduced following the split-dose protocol. Painter (134) has shown that an initial low UV dose alters the pattern of DNA synthesis in cell cultures, causing an abnormal distribution of the sizes of newly synthesized DNA at the time of a second UV dose. Consequently, the larger size of the pulse-labeled DNA extracted from cells exposed twice to UV light might result from preferential labeling of molecules of larger size during the pulse. However, this criticism has been rebutted by the observation that if the first UV dose is given in the G_2 phase of the cell cycle, enhanced elongation of newly replicated DNA still occurs in cells irradiated with a second UV dose prior to S phase (133). Since DNA replication does not occur between these two doses, this enhancement cannot be attributed to altered patterns of DNA synthesis after the first UV exposure.

Difficulties still remain in interpreting these results (133). Since the enhancement is a small one, it is not known whether it is a major or minor mechanism for facilitating DNA replication of UV-irradiated templates. Also, since DNA is isolated from the entire irradiated cell population, it is not known whether this enhanced elongation of nascent DNA strands occurs in viable cells, and might, therefore, be a mechanism for cell survival. Finally, the effect is observed in cells receiving both UV doses prior to entering the S phase. As the cells tested had not attempted to replicate their damaged DNA and were also deficient in dimer excision, it is not clear how they could be induced to express a new replication mechanism.

Evidence that DNA replication may be enhanced by an inducible mechanism is also suggested by delayed analyses of the DNA from UV-irradiated cells (135–137). As described previously (Section III,E), pulse-labeling shortly after irradiation (30–120 minutes) yields DNA smaller than the DNA from unirradiated cells, suggesting that replication is inhibited on irradiated templates. However, if these same irradiated cells were pulse-labeled for five or more hours after irradiation, the DNA had a size approaching that of DNA from unirradiated cells. This observation has been made in mouse cells (*135*), human fibroblasts including XP cells (*136*), and hamster cells (*137*). Since these rodent and XP cells were deficient in dimer excision (*21*), this late synthesis of normal-sized DNA could not have resulted from replication of repaired, dimer-free templates. Therefore, it might be concluded that a new mechanism of DNA replication facilitating replication of dimer-containing chromosomes was induced in these cells. As discussed for bacteria (Section II,E), this enhanced elongation might arise either from rapid removal of gaps in the newly replicated DNA, or from continuous synthesis past pyrimidine dimers.

However, there may be an alternative explanation for the above observations. In most cases, unsynchronized cells were used, and the time between the end of one S phase and the beginning of the next was not indicated. Therefore, larger DNA might have originated from a new round of DNA synthesis on damage-free templates previously synthesized on irradiated templates. Unlike bacteria, in which pyrimidine dimers are rapidly dispersed between parental and newly replicated strands, dimer transfer in mammalian cells occurs only at a low level if at all, so that dimer-free template strands might be formed following completion of one round of replication. Consequently, these results are not convincing evidence that mammalian cells acquire the ability to synthesize continuous DNA on UV-damaged templates. More convincing evidence for an inducible repair mechanism in mammalian cells that enhances repair of animal viruses is discussed in Sections V,C and V,D.

IV. Mutagenesis by Ultraviolet Radiation

A. Mutagenesis in Bacteria

Ultraviolet light is mutagenic for bacteria. When UV-irradiated bacteria are subsequently treated with photoreactivating light, the frequency of mutants per survivor decreases substantially. Thus, pyrimidine dimers are considered to be responsible for much of the observed mutagenesis (see *13*, *17*, *137a*). The observation that UV is mutagenic and also causes DNA damage suggests that UV-induced mutations might be caused by a DNA-repair pathway or a process involved in tolerance of DNA damage. However, it is possible that mutagenesis is not causally related to a repair process at all, but is associated with repair because both processes occur as a consequence of UV damage.

1. THE SOS HYPOTHESIS

Mutagenesis is one of a variety of physiological changes that occur in UV-irradiated bacteria (17). These changes include: (a) increased synthesis of recA protein (see Section II,H) and probably increased or decreased synthesis of other proteins as well; (b) induction of a resident prophage; (c) resistance to DNA degradation following further DNA damage; (d) increased ability to repair UV-irradiated bacteriophage lambda and certain other bacteriophage strains (designated Weigle or UV reactivation; see Section V,A); (e) increased mutagenesis of UV-irradiated bacteriophage lambda and other bacteriophage strains (Weigle mutagenesis; see Section V,E); (f) reinitiation of DNA synthesis by an abnormal mechanism (see Section II,G); and (g) a temporary inhibition of cell division. The expression of all these physiological changes depends principally upon normal expression of the *lexA* and *recA* genes, although expression of a number of other genes is required as well (reviewed in 17). These functions share the common feature that protein synthesis after UV irradiation is probably necessary for their efficient expression. In addition, many treatments that inhibit DNA synthesis usually lead to expression of the entire set of functions. The "SOS hypothesis" proposes that each of these UV-induced cellular changes is part of a system of adaptation necessary for optimal survival following DNA damage. It has been further proposed that SOS functions are not normally expressed, but remain under the regulatory control of the *lexA* and *recA* genes until there is interference with DNA synthesis.

2. MUTAGENESIS IN REPAIR-DEFICIENT MUTANTS

In bacteria, mutagenesis by UV is genetically controlled. Mutations in the *recA* and *lexA* genes (and in several other genes) prevent UV-induction of mutations in *E. coli* (17). In contrast, mutants deficient in normal excision repair, i.e., uvr^- or $polA^-$ (deficient in DNA polymerase I) mutants (see Section I,B), show abnormally high rates of UV-induced mutagenesis at low UV doses (17). These observations suggest that DNA synthesis associated with excision repair is highly accurate, but that other DNA repair or tolerance mechanisms may be error prone.

Studies with recombination deficient ($recB^-$ $recF^-$) mutants of *E. coli* show that UV-induced mutagenesis may be obtained in the nearly total absence of genetic recombination. The *recB* and *recF* genes control the only two known pathways of genetic recombination in *E. coli;* the *recA* gene must also function for both of these pathways to be expressed (*138*). The *recB*⁻ and *recF*⁻ mutants are deficient in postreplication repair of UV damage, a process that involves frequent genetic exchanges in wild-type cells (see Section II,D). The properties of these mutants, therefore, suggest that mutagenesis is associated with a mechanism other than postreplication repair or genetic recombination, but they do not eliminate the possibility that some mutagenesis might be associated with a postreplication repair or recombination mechanism in wild-type cells.

3. Mutagenesis in Derivatives of *lexA* and *recA* Mutants

Since *lexA* appears to be a repressor of *recA* (see Section II,H), it might appear that both these gene products control the same pathway of UV-induced mutagenesis. However, genetic analyses of these mutants suggest a more complex regulation of mutagenesis. These analyses distinguish three possible ways in which *lexA* might control UV-induced mutagenesis: (*a*) regulation of *recA;* (*b*) direct production of mutations; and (*c*) regulation of other genes whose products are required for mutagenesis.

It appears that a high level of expression of *recA* protein is not sufficient for UV-induced mutagenesis. This result suggests that regulation of *recA* by *lexA* alone does not control UV-induced mutagenesis (mechanism a). As described in Section II,H, *lexA* protein may be a simple repressor of *recA* [see Note Added in Proof (b)]. After induction, *recA* becomes active as a protease and cleaves the *lexA* repressor, thereby inducing its own synthesis. *lexA*⁻ mutants are thought to make a repressor refractory to the *recA* protease, so that induction of *recA* protein does not occur in these strains. Strains carrying mutations believed to lie in the operator region of the *recA* gene have been isolated (*79, 80*) (see Section II,H). In cells that are either *lexA*⁺ or *lexA*⁻, these *recA*-operator mutations cause a high, constitutive level of synthesis of *recA* protein, indicating the failure of *lexA* to repress *recA* in these strains (*79, 80*). Following irradiation of double mutants carrying *lexA*⁻ and *recA*-operator mutations, UV-induced mutagenesis is not observed (*79, 80*). Consequently, mutagenesis requires an additional factor other than high levels of *recA* protein.

Ultraviolet-induced-mutagenesis is observed in *E. coli* strains carrying amber and missense mutations in *lexA* (*138*), indicating that the *lexA* gene product is dispensable for mutagenesis (mechanism b) (*139, 140;* L. Z. Pacelli and D. W. Mount, unpublished observations). However, *lexA*⁺ appears to regulate mutagenesis, as shown by the observation that, although $lexA^-$ $recA$-operator mutants fail to exhibit UV-induced mutagenesis, $lexA^+$ $recA$-operator mutants do (79, 80). The protease activity associated with $recA$ in $lexA^+$ mutants inactivates the $lexA$ protein, while $lexA^-$ protein is resistant to $recA$ protease cleavage. These results suggest that $lexA$ protein may repress the expression of some genes (other than $recA$) required for UV-induced mutagenesis (mechanism c), and that $recA$ cleavage of the $lexA$ protein may be required for expression of these other gene products.

The behavior of $lexA^-$ $recA$-operator mutants just described suggests that one function of the $recA$ product in mutagenesis is to cleave $lexA$ protein. If this cleavage were the only function of $recA$ in mutagenesis, introducing a missense or amber mutation into the $lexA$ gene to inactivate its function should make mutagenesis independent of $recA$. Several such mutant strains have been constructed and have been found to be nonmutable by UV (D. Mount, unpublished observations). We conclude that the $recA$ gene product serves at least one function in UV-induced mutagenesis other than inactivation of the $lexA$ protein.

There is another type of mutation in the $recA$ structural gene (tif) that influences mutagenesis (17). tif mutant cells constitutively express high levels of the tif mutant form of the $recA$ protein in the absence of an inducing treatment when grown at 42°C (45, 82, 84, 85). Even higher levels of the tif mutant form of $recA$ are observed when the tif protein is derepressed by an additional mutation destroying $lexA$ function (45, 82, 84, 139). Under these conditions, the cells exhibit abnormally high levels of spontaneous mutagenesis even in the absence of DNA damage, and produce additional mutations when UV-irradiated (17, 140). The origin and nature of the spontaneous mutations that arise in tif mutant cells have not been determined. One possibility is that they arise through mutagenesis associated with the repair of spontaneously occurring DNA lesions by a mechanism similar to that induced by UV. Another is that the tif mutant form of the $recA$ protein binds to the chromosome and modifies its structure so that replication is more inaccurate. A third possibility is that the proteolytic activity of this protein could become active and modify proteins that affect replication fidelity.

The existence of the tif mutant implies that the $recA$ protein must be activated in the induced cell in order to perform its mutagenic function. According to this hypothesis, the tif mutant form is spontaneously activated, in the absence of any inducing treatment (45, 140). More direct evidence for a required activation of $recA$ protein is provided by the properties of mutants that overproduce the normal form of $recA$ protein. Strains carrying either an amber mutation in the $lexA$

gene or a *recA*-operator mutation must be given an inducing treatment (such as UV irradiation) in order to express most SOS functions (79, 80, 139, 140). These strains do not show a high level of spontaneous mutations, as do *tif* mutants, but they are capable of UV-induced mutagenesis. Therefore, activation of *recA* protein, presumably to some form that can influence replication fidelity, is required for UV mutagenesis.

The experiments described above suggest the following model for UV-induced mutagenesis: (a) cleavage of the *lexA* protein by the *recA* protease to derepress *recA* and at least one other gene under *lexA* regulation; and (b) activation of *recA* protein for one or more additional functions in mutagenesis. These functions have not yet been identified, but probably involve a combination of the DNA binding and protease activities of the *recA* protein. The most straightforward explanation for the role of *recA* protein in mutagenesis is that it modifies either the structure of the cell chromosome, by binding to it, or induces or modifies an enzyme required for DNA replication so that the fidelity of replication is relaxed.

The mechanism whereby UV-induced mutations are produced has not yet been elucidated in bacteria. However, the evidence suggests two possibilities. First, analysis of UV-induced mutations (see below) shows that many of these mutations are produced opposite pyrimidine pairs in the template strands, suggesting that they might have been produced by some type of DNA synthesis past pyrimidine dimers. Since normal base-pairing at pyrimidine dimers is disrupted, such replication may involve insertion of one or two incorrect bases opposite the damaged sites.

An alternative mechanism for the induction of mutations following UV irradiation comes from studies of mutagenesis of bacteriophage lambda (Section V,B). Mutations in this phage appear to be produced not only in damaged phase DNA, but also in undamaged phage, provided that the host cell has been irradiated to induce functions required for mutagenesis. This result suggests that inducible host functions might relax the fidelity of replication and thus produce mutations during replication of undamaged DNA (so-called "untargeted mutagenesis"). It is not known whether such a mechanism might also act to produce mutations in *E. coli* DNA.

4. Tandem Base-Pair Changes Result from UV-Induced Mutagenesis

Coulondre and Miller (141) analyzed the spectrum of base changes resulting from forward mutations induced by UV light in the *lacI* gene of *E. coli*. They isolated over 5000 nonsense mutations, employing a variety of mutagens. Through a fine-structure genetic analysis of these mutations, they could assign many of them to specific codons and thereby deduce the precise base changes that had occurred. A significant percentage (approximately 2%) of the UV-induced nonsense mutations were tandem double-nucleotide changes. Nearly all of these tandem base-pair mutations occurred at the positions occupied by two adjacent pyrimidines in the wild-type codon, as expected if incorrect nucleotides had been inserted during replication past pyrimidine dimers or during repair replication. Furthermore, this analysis revealed that of several mutagens tested only UV produced these tandem, double-nucleotide changes.

Estimates of the actual percentage of double mutations in these experiments suggested that only a fraction of all the mutations produced was being detected experimentally. First, double base-pair changes in many codons would not have given rise to nonsense mutations, because of the nature of the genetic code. Second, the method of analysis may have incorrectly identified many tandem double-nucleotide changes as single-nucleotide changes, becuse the second mutant nucleotide was in an adjacent codon. Third, it is also possible that some of the single base-pair changes detected as nonsense mutations represented the chance incorporation of one correct and one incorrect nucleotide opposite a pyrimidine dimer. These considerations suggest that a substantial fraction (much greater than 2%) of UV-induced mutations are of the random double-base-pair type and may, therefore, have been induced during replication past pyrimidine dimers.

5. Other Types of UV-Induced Mutations

Several experiments reveal that UV irradiation also produces mutations in *E. coli* other than simple base substitutions. It probably induces as many deletion mutations as base-substitution mutations, and the production of these deletions depends upon the function of the *recA* gene (*142*). UV can induce histidine auxotrophs of *Salmonella typhimurium* whose reversion pattern by other mutagens suggests that they are frame-shift mutations (*143*). Since these observations were made, translocatable genetic elements (transposons) have been identified (*144*). Deletions, inversions, and frame-shift mutations are associated with the insertion and excision of these elements. It is therefore possible that the deletions and frame-shift mutations detected following UV irradiations might be associated with transposon movement. However, chromosome rearrangements associated with movement of one transposon, tn10, in normally growing cells are *recA*-independent

(*145*). Since *recA* mutants fail to exhibit UV-induced mutagenesis, a role for transposons in such mutagenesis is still purely hypothetical.

B. Mutagenesis in Mammalian Cells

As in bacteria, the survivors of UV-irradiated mammalian cell cultures contain more mutations than do cultures of unirradiated cells (*146, 147*). Since all repair-deficient mammalian cell lines tested exhibit UV-induced mutagenesis, the mammalian cell functions required for mutagenesis are not known. However, comparisons of normal and xeroderma pigmentosum human cells lines have been useful in evaluating the requirement for cellular gene products in UV-induced mutagenesis.

In bacteria, UV-induced mutations are believed to result from replication of DNA containing pyrimidine dimers (*17*). Excision repair in these cells is believed to be an accurate process (Section IV,A). Likewise, in mammalian cells, mutations are thought to be produced by replication of UV-damaged DNA, not by excision-repair synthesis. However, there is presently no information as to whether mutations are produced at sites of UV damage, or whether mutations can also occur during replication of undamaged DNA in irradiated mammalian cells.

Evidence that excision repair does not produce mutations in mammalian cells comes from experiments in which nonreplicating monolayer cultures of human fibroblasts were irradiated with UV light (*148*). The frequencies of mutants among the survivors at a given dose were determined both for cultures that were replated and allowed to undergo cell division immediately after UV treatment and for cultures maintained as nondividing, monolayer cultures for 7 days after irradiation prior to replating. In normal human fibroblasts, the frequency of mutations was high in cultures replated immediately, and much lower in cultures maintained for a week as a monolayer before replating. During this incubation time, cells presumably were able to repair by excision most of the DNA damage that might have induced mutations. Consequently, excision of pyrimidine dimers appears to be an accurate process in these cells. In constrast, XP cells gave comparable numbers of mutants whether monolayer cultures were replated immediately or after 7 days of incubation. Since these cells were almost totally deficient in the ability to excise pyrimidine dimers, it appears likely that unexcised pyrimidine dimers present in cellular DNA at the time of cell division may have been responsible for the production of mutations. However, the cell lines used in these studies were not isogenic. Therefore, it cannot be concluded with certainty that the observed effects on mutagenesis were due to the XP defect.

Excision repair has also been shown to be an accurate repair mechanism by comparing the frequency of mutant survivors in irradiated cultures of normal and XP human fibroblasts (149–151). The frequency of mutation at each UV dose is higher for XP cells than for normal cells. However, if the mutant frequency was plotted as a function of the percentage survival, the results obtained for XP cells are comparable to those obtained with normal cells. Consequently, the extent of DNA repair necessary to achieve a given survival level in the presence or absence of excision repair produced an equal yield of induced mutations. Because normal levels of mutagenesis were obtained in the absence of functioning excision repair mechanisms in human cells, it may be concluded that mutagenesis is associated with repair mechanisms other than excision repair in these cells.

Results of similar experiments with human XP variant cells have suggested that these cells may be deficient in one or more pathways that allow accurate replication of DNA containing UV damage. At a given level of survival, the XP variant cell line XP4BE showed a higher fraction of cells containing a mutation in the guanine phosphoribosyltransferase gene than did either normal or excision-deficient XP cells (152, 153). At present this observation has not been extended to other XP variant cell lines or to other mutational systems.

The defective DNA repair function in XP variant cells has not been identified. These mutant cells appear to be deficient in replication of DNA after irradiation and may, therefore, be deficient in an error-free replication pathway. Gradient analysis of pulse-labeled DNA from irradiated XP variant cultures indicate that nascent DNA from these cells is shorter than DNA from either the same, unirradiated mutant cells or normal cells irradiated with the same UV dose (56, 57, 107; see Section III,F). If the pulse-labeled XP cells were further incubated for several hours in unlabeled medium, the size of the labeled DNA molecules increased at a rate comparable to that found in normal cells (56, 107), eventually reaching a size slightly smaller than the DNA of unirradiated cells. These experiments suggest that replication forks in XP variant cells may be more severely blocked at sites of UV damage than in wild-type cells, but that replication eventually proceeds past these lesions. This same mechanism could also increase the chance that mutations are produced.

A mechanism for enhanced survival of UV-irradiated bacterial cells or bacteriophage that appears to be associated with the production of mutations and the induction of the $recA^+$ gene product has been observed (see Sections IV,A, V,A, V,B). Experiments designed to test whether a similar inducible mechanism might act to produce mutations in mammalian cells have been negative (132, 154). In these experiments, mutagenesis of hamster cells irradiated with a single large dose of UV was compared to that of cells treated first with a low dose of UV or X-rays followed by an incubation period and treatment with a second, larger UV dose. No increase in mutagenesis following the split-dose protocol was observed. In contrast to these results, mutagenesis of animal viruses (see Section V,D) may proceed by a mechanism that requires DNA damage to induce the host cell prior to infection.

V. Mechanism for Reactivation of Ultraviolet-Damaged Viruses

A. Repair of Bacteriophage

The availability of well developed genetic systems employing both viral and host mutants has provided a means for dissecting the DNA repair pathways in bacteria infected with UV-irradiated bacteriophages. The types of DNA repair identified are host-cell reactivation, multiplicity reactivation, prophage reactivation, and UV reactivation (155).

Host-cell reactivation is brought about by excision of pyrimidine dimers from UV-damaged bacteriophages (such as lambda) by the host-cell excision system, and probably does not involve viral functions. This type of reactivation is demonstrated by the reduced ability of UV-irradiated lambda (and other phages) to form plaques on excision-deficient host mutants (155).

Multiplicity reactivation and prophage reactivation depend upon the presence of homologous bacteriophage DNA in the cell and a functional system of recombination. In the case of multiplicity reactivation, cells are multiply infected with irradiated bacteriophage, mixed with an excess of unirradiated, indicator bacteria, and incubated to allow plaques to form. By this assay, many more cells multiply infected by irradiated bacteriophage produce active virus than do single infected cells. This increased production of reactivated virus is termed multiplicity reactivation. In cells infected by bacteriophage lambda, the virus and the host provide independent recombination functions, at least one of which must be functional for multiplicity reactivation to occur; these functions are deficient in host $recA^-$ and viral red^- mutants, respectively (155, 156). Multiplicity reactivation is considered to arise by genetic recombination between the damaged viruses in the multiply infected cells. This mechanism could involve breakage and joining of damaged homologous chromosomes to produce genomes free of damage.

MECHANISMS OF DNA REPLICATION AND MUTAGENESIS 109

Prophage reactivation is observed when unirradiated, lysogenic cells are infected with UV-irradiated bacteriophage (such as lambda), homologous to the prophage (155). The prophage carries mutations that prevent its own replication but allow normal multiplication of the superinfecting virus. The prophage thus provides homologous, damage-free DNA with which the incoming damaged virus can recombine. The yield of active virus from these cells is greater than that from a similar nonlysogenic host. Prophage reactivation depends upon a functioning system for recombination, as does multiplicity reactivation. Accordingly, the presence of the prophage probably promotes survival of the UV-irradiated, superinfecting bacteriophage lambda by providing short, undamaged segments of homologous DNA for insertion into the damaged chromosome of the superinfecting virus.

UV reactivation (also called Weigle reactivation) occurs when UV-irradiated cells are singly infected with UV-irradiated bacteriophage (17, 155). Virus production by these cells is then monitored by a plaque assay, as in the multiplicity reactivation protocol. Higher survival of irradiated phage is observed with cells previously UV-irradiated than with untreated cells. The extent of UV reactivation depends upon the dose given to the host cell and also upon the genotype of the host cell. Typically, virus survival can be increased from a value of 1%, when unirradiated cells are infected, to 10%, if cells treated with an appropriate dose of UV are employed. Protein synthesis must also take place in the irradiated cells prior to infection if reactivation is to occur. Thus, it appears that the inducing treatment given the host cells enhances the synthesis of one or more proteins that promote repair of the virus. Introduction of damaged DNA into the cells by conjugation (157) can also promote enhanced repair of irradiated lambda.

Mutations in the *lexA* and *recA* genes prevent UV reactivation. Because of this genetic requirement and the requirement for protein synthesis, UV reactivation may be a manifestation of a cellular SOS function (17). *tif* mutants and related mutants that spontaneously express various SOS functions constitutively, i.e., in the absence of DNA damage, also show spontaneous expression of the UV reactivation pathway (17, 140, 158). Thus, UV reactivation occurs under the same conditions as spontaneous cellular mutagenesis in these mutants (Section IV,A), suggesting a relationship between these two processes.

One plausible mechanism for the enhanced virus survival in UV-treated cells is that the high levels of *recA* protein in the infected cells permit reconstruction of partially replicated, UV-damaged chromosomes containing gaps opposite dimers by a recombination mechanism involving sister chromosomes, as occurs in uninfected, irradiated bacteria. A second possible mechanism is increased efficiency of excision repair in excision-proficient cells. A third mechanism could be one that allows insertion of nucleotides opposite pyrimidine dimers, a process that should be mutagenic.

Ultraviolet reactivation of both single-stranded (ϕX174, S13) and double-stranded (λ) DNA viruses has been observed (155, 159, 159a). The efficiency of UV reactivation of ϕX174 is somewhat lower than that of phage lambda. This observation may result from an inability of ϕX174 to utilize either recombination or excision repair during single infections, whereas lambda should be repaired by both mechanisms under conditions of a single infection.

The UV reactivation of phage lambda is deficient in an *E. coli* mutant designated *umuC* (159b), whereas other SOS functions, such as prophage induction, appear to be expressed normally in this strain. The *umuC* function, therefore, seems to be necessary to promote specific expression of the UV-reactivation function. UV-induced mutagenesis of the host is also deficient in *umuC* mutants, indicating that the type(s) of UV reactivation promoted by *umuC* is (are) associated with mutagenesis.

B. Mutagenesis of Ultraviolet-Irradiated Bacteriophage

Of the various repair mechanisms available to UV-irradiated bacteriophage lambda, mutagenesis of the damaged bacteriophage is associated only with UV reactivation (155). As described (Section IV,A) for mutagenesis of cellular DNA, bacteriophage mutagenesis may also require inducible, cellular functions and may result from errors produced during viral replication, although other mechanisms have not been excluded.

To observe this mutagenesis, unirradiated or irradiated cells are singly infected with unirradiated or irradiated bacteriophage lambda, following which the infected cells are assayed for plaque-forming ability with an excess of unirradiated bacteria (155). Without irradiation treatment of host or virus, most plaques produced have a turbid center from lysogenization of the indicator bacteria. Approximately 1 in 2000 plaques has a clear center because it arose from a mutant bacteriophage that did not produce active repressor and was therefore unable to lysogenize the host cells. If both the virus and the host are irradiated with UV light, the mutant yield increases up to 10-fold, depending upon the UV doses given to the cells and virus. No increase in mutagenesis is observed (*a*) in irradiated $recA^-$ or $lexA^-$ mutants; (b) in unirradiated cells; or (c) in irradiated cells in which protein synthesis is inhibited prior to virus infection. These genetic and physiological requirements are similar to those needed for cellular, UV-induced mutagenesis and UV reactivation of damaged virus (17). Consequently, these processes might result from the activity of the same mutagenic repair mechanism, although additional experiments are required to establish this conclusion firmly.

Further analysis of this system reveals that phage mutants can arise by two, possibly independent, mechanisms. Single-burst analyses show that the clear plaques arising on the strains of indicator bacteria usually employed are pure mutant clones; i.e., essentially all the bacteriophage produced in these clear plaques carry the repressor mutation (160). These pure mutant clones appear to arise from the production of mutational homozygotes at a very early stage of virus replication. One possible mechanism whereby such mutants might arise is that an inducible host function that repairs or replicates damaged DNA with a high probability of error might rescue heavily damaged phage unable to complete the first round of replication.

Approximately 10-fold more infected cells produced clear plaque, mutant phage than were detected by the above method (160). These additional mutants were present in mixed bursts of mutant and nonmutant virus and were not detected because of complementation of mutants by wild-type virus. Mixed bursts presumably arose from mutational heterozygotes that segregated during replication to form wild-type and mutant genomes. To produce pure, mutant clones, both the virus and host cell must have been irradiated by UV light. In contrast, mixed bursts of wild-type and mutant viruses were produced when UV-irradiated host cells were infected with either irradiated or unirradiated bacteriophage, although a higher frequency of mutant clones was obtained using UV-damaged virus.

These differences in the physiological requirements for detection of pure and mixed clones of viral mutants suggest that two different mechanisms may be involved in the production of these mutants. Possibly, irradiation of the host cell increased the chance of errors during replication of undamaged virus by inducing a host function that relaxed the fidelity of replication. Irradiation of the virus, in addition, might have produced additional mutants either by replication past sites of DNA damage, or during repair of these lesions. Pure mutant clones might be produced by only this latter repair mechanism, whereas mixed bursts might be derived from both mechanisms. One repair-deficient mutant of *E. coli* (*recF*⁻) has been found to inhibit the production of pure, but not of mixed, mutant clones (*161*), whereas, in repair-deficient *lexA*⁻ and *recA*⁻ mutants, the production of both types is reduced (*160*). The similarity of these genetic requirements for both host and viral mutagenesis suggests that viral and cellular mutagenesis may occur by similar biochemical pathways.

Mutagenesis associated with UV reactivation of single-stranded viruses, such as S13 and ϕX174 (*159, 161a*), provides additional evidence that replication infidelity may be the mechanism of production of many UV-induced mutations, and that inducible host functions may be required. First, studies of UV reactivation suggest that the increased survival of these single-stranded phages arises from enhanced replication of viral genomes (see Section V,A). This type of DNA synthesis might produce mutations by insertion of incorrect nucleotides opposite pyrimidine dimers. Second, the host cells must have been previously irradiated for successful mutagenesis of the infecting, UV-damaged phage. Consequently, an inducible mechanism may be involved. Additional evidence that this inducible mechanism requires host-cell functions has been suggested by the observations that (*a*) these viruses have a limited coding capacity and probably do not contribute mutagenesis functions; and (*b*) viral mutagenesis depends upon the host *lexA* and *recA* functions (*17*).

Mutagenesis of phage S13 also depends upon the host *recF* function, as does the production of full mutant bursts of lambda described above (A. J. Clark, personal communication). The *recF* product is therefore strongly associated with the induction of mutations in heavily damaged double-stranded or damaged single-stranded DNA. We anticipate that the type of mutagenesis controlled by *recF* should occur at the site of the lesion, in contrast to untargeted mutagenesis, which should occur in cells that have altered replication fidelity.

C. Repair of Animal Viruses

As is the case for UV-damaged bacteriophages (Section V,A), animal viruses can be repaired after exposure to UV by at least three mechanisms: host-cell reactivation, multiplicity reactivation, and UV reactivation. These terms have been borrowed from the bacteriophage literature because the repair mechanisms appear to resemble closely those for damaged bacteriophages.

Host-cell reactivation of SV40 (*47*), human adenovirus 2 (*48*), and herpes simplex virus type 1 (*46, 49*) has been observed by comparing the relative survival of UV-irradiated virus in normal human cells and in XP cells deficient in excision repair. When these cells were infected with UV irradiated virus in a plaque assay, the survival of plaque formation was lower for the repair-deficient cells than for normal cells. Consequently, as is the case for bacteriophage, host-cell excision enzymes appear to repair damaged viral DNA in excision-proficient, normal cells.

Multiplicity reactivation has also been reported for several animal viruses, including herpes simplex virus type 1 (*123*), SV40 virus and human adenovirus 12 (*122*). The experimental protocol with herpes simplex virus was similar to that used to detect multiplicity reactivation in bacteriophage-infected bacteria. Human cells were either singly or multiply infected with UV-irradiated viruses, and the ability of these cells to form a plaque when seeded with an excess of uninfected cells was determined. For singly infected cells, a very low ($< 10^{-2}$ plaque-forming units per cell) multiplicity of infection was used, because unirradiated animal virus stocks contain defective viruses (i.e., virus particles containing incomplete genomes and unable to form plaques), which are typically present in an approximate 100-fold excess over plaque-forming particles. The results obtained indicate that fewer singly infected cells produced active virus at a given UV dose then multiply infected cells, suggesting that viral repair in multiply infected cells may involve mechanisms not active in singly infected cells. As discussed in Section III,I, the observation that UV stimulates genetic exchanges between viral genomes in multiply infected cells favors the interpretation that multiplicity reactivation of animal viruses involves genetic exchanges between damaged viruses.

There is little evidence to indicate which host or viral genes might be involved in multiplicity reactivation. Experiments with herpes simplex virus type 1 have shown that multiplicity reactivation is not impaired in either excision-deficient XP cells or in XP variant cells (*120*). These results suggest either that the deficient functions in these cells are not required for multiplicity reactivation of the virus, or that herpes simplex virus codes for similar functions. At least one cellular function, however, appears to be necessary for multiplicity reactivation of herpes simplex virus, since multiplicity reactivation of this virus was reduced in the Bloom's syndrome cell line, CRL 1492 (*123*). Other Bloom's syndrome lines are capable of nearly normal levels of multiplicity reactivation of herpes virus

(123). The chromosomes in Bloom's syndrome cells have been found to contain extremely high levels of spontaneous sister chromatid exchanges (162), as if these cells either contained a hyperactive recombination function or, as might explain the deficiency in multiplicity reactivation, were unable to complete recombination events normally. Since alternative possibilities for the defect in Bloom's syndrome cells also exist, it is not possible, at the present time, to anticipate how the deficient cellular function(s) in these cells might affect multiplicity reactivation of herpes simplex virus.

Ultraviolet reactivation has also been reported for several animal viruses, including herpes simplex virus type 1 (163), human adenovirus 2 (164), and SV40 (165, 166). These experiments were performed using a protocol similar to that employed to study UV reactivation in bacteriophage-infected bacteria. Prior to infection, cells (human or monkey) were treated with a DNA-damaging agent such as UV, ionizing radiation, or chemical carcinogens [e.g., metabolically activated aflatoxin B1, 2-(N-acetoxy)acetamidofluorene, methyl methanesulfonate, ethyl methanesulfonate]. Cells were incubated prior to infection and subsequently infected in a plaque assay at a low multiplicity of infection. Treated cells produced more plaques following infection with irradiated virus than untreated cells. Furthermore, in experiments with herpes simplex virus type 1, protein synthesis was required after treatment of the cells, since UV reactivation was not observed if cells were incubated in the presence of cycloheximide prior to infection (167, 168). These results suggest that the DNA-damaging treatment to the host cells induces a cellular repair mechanism that can then repair UV-irradiated virus more efficiently than repair mechanisms present in untreated cells.

D. Mutagenesis Associated with Ultraviolet Reactivation of Animal Viruses

As described below, UV irradiation of mammalian host cells is associated with mutagenesis of infecting viruses. It is premature to state that such mutagenesis might arise from altered replication fidelity, as may occur in bacteriophage-infected *E. coli*. However, some evidence in favor of such a mechanism exists. There is also no evidence to indicate whether viral and cellular mutagenesis occur by similar mechanisms. In SV40-infected monkey cells, reversion of two different viral temperature-sensitive mutations (one in gene A required for viral replication and one in a coat protein gene) occur more readily in UV-irradiated than in unirradiated host cells (169). As is the case for UV reactivation, incubation of the irradiated host cells was necessary prior to infection to observe this increased mutagenesis. These results suggest, therefore, that the UV-damaging treatment to the host cell might induce a cellular pathway that allows UV-damaged viruses to be replicated or repaired with reduced accuracy.

These studies (169) also provide evidence that increased mutagenesis of viruses in irradiated cells does not arise from mechanisms involving recombination and may, therefore, be due to reduced fidelity of viral replication, although other mechanisms, such as reduced fidelity of excision repair, have not at present been eliminated. In these experiments, conditions were maintained so that cells were never infected with more than one viral particle during induction of mutations. Therefore, repair mechanisms involving multiple viral genomes (e.g., multiplicity reactivation) could not be responsible for this UV-induced mutagenesis. However, recombination between homologous regions of sister chromosomes produced by replications of single SV40 genome cannot be eliminated as a source of mutations. Since animal virus stocks typically contain defective viruses in vast excess (100-fold higher or greater) over plaque-forming particles (170–172), the possibility of recombination between plaque-forming genomes and defective genomes must be considered as a source of mutations in these experiments. However, the low multiplicity of infection employed in these experiments greatly reduces this possibility. Furthermore, stocks of the temperature-sensitive virus should contain defective virus particles, which also carry the mutant allele. Consequently, these stocks should not be a source of base substitutions that could allow formation of a temperature-resistant mutant virus by a recombination mechanism. Finally, restriction enzyme analysis of these revertant viruses failed to detect any deletions or insertions in the DNA of these viruses greater than 10 bases, the limit of accuracy of the analysis. Consequently, large rearrangements of the viral genome do not appear to be involved in the production of viral mutations in UV-irradiated host cells.

Virus-infected mammalian cells, such as those described above, provide an ideal system for asking whether viral mutations produced following infection of UV-irradiated host cells might arise from inaccuracies during replication of the viral genomes and whether these mutations might occur by insertion of incorrect bases opposite sites of pyrimidine dimers. Studies of mutations that follow infection by single-stranded viruses would have the further advantage of avoiding any contribution of recombination to mutagenesis, since no complementary strand exists in virus-infected cells prior to replication. On the other hand, although SV40 is a double-stranded virus, the entire sequence of the genome is known (*173*), thus facilitating the analysis of the mutations induced. We advocate the use of such viral experiments to elucidate mechanisms of mutagenesis in mammalian cells.

Experiments comparable to those described above for SV40 indicate that both unirradiated and UV-irradiated herpes simplex virus grown on UV-irradiated monkey cells contain more mutations in the viral thymidine kinase gene than do viruses grown on unirradiated cells (*167, 174*). However, this increase in mutagenesis was observed only when the host cells were infected at a high multiplicity of infection (one plaque-forming unit or more per cell) and was not observed at a multiplicity of infection of 0.2 (*174*), suggesting that the increased mutagenesis observed in UV-irradiated cells might involve genetic recombination. Under these conditions, it is likely that many defective, non-plaque-forming particles would also have been present in the initially infected cells, since defective particles containing incomplete genomes (including deletions, duplications, and other abnormal DNA sequences) exist in all herpes stocks (*175*). Consequently, it is possible that defective viruses present in the herpes stocks used for these experiments might have contained extensive deletions in the thymidine kinase gene. Subsequently, recombination between these defective particles and plaque-forming particles might have produced plaque-forming particles with deletions in the thymidine kinase gene. If recombination was stimulated in host cells treated by UV, such recombination might be responsible for the apparent increase in mutation frequencies in UV-irradiated host cells. Further characterization of the thymidine kinase mutations induced under these conditions should indicate whether this mechanism might account for the results obtained.

VI. Effects of Ultraviolet Irradiation on DNA Synthesis in Vitro

Analysis of DNA synthesis in UV-irradiated cells is made difficult by the complex cellular regulatory mechanisms that influence various steps in the replication process. Studies of *in vitro* replication using UV-damaged templates have allowed investigation of this process in the absence of such complications. These *in vitro* studies have also allowed analysis of specific components of the replication complex with regard to their influence on the replication of UV-damaged templates. As described below, these studies have served as models for DNA replication *in vivo* and have provided substantial information concerning the mechanisms of DNA replication and mutagenesis utilizing damaged templates.

A. Replication of Ultraviolet-Damaged DNA

Synthesis of DNA *in vitro* by DNA polymerases from prokaryotic cells is severely inhibited by UV irradiation of the DNA template. In contrast, the response of eukaryotic polymerases to a similar treatment of template DNA is, at present, a source of controversy.

A comparison was made of the ability of DNA polymerases from both prokaryotic (polymerase I and III of *Escherichia coli*) and eukaryotic (polymerase alpha of calf thymus) cells to replicate UV-irradiated DNA from the single-stranded bacteriophage ϕX174 using a complementary restriction fragment as a primer (176). While the rate of incorporation of deoxyribonucleoside triphosphates into DNA synthesized by the *E. coli* polymerases was considerably reduced by UV irradiation of the template DNA, synthesis promoted by the mammalian polymerase was much more resistant to template irradiation. These results suggest that prokaryotic polymerases might be inhibited at sites of UV damage, while eukaryotic polymerases might be capable of replication past such lesions.

A possible explanation for this difference might be the presence of the 3'-to-5' exonuclease associated with prokaryotic DNA polymerases, but not present in the eukaryotic polymerase (72, 177). Since this nuclease is thought to have a "proofreading" function during replication *in vivo* (178), it has been proposed that the *E. coli* DNA polymerases insert nucleotides opposite pyrimidine dimers, but then remove these nucleotides using this "editing" nuclease. Mammalian polymerases, in contrast, might insert bases but not remove them. Supporting this idea is the observation of a high turnover rate of deoxyribonucleoside triphosphates to monophosphates in the reaction mixtures described above, using prokaryotic DNA polymerases (176). This result is consistent with a model in which polymerases repeatedly insert and remove nucleotides opposite the sites of pyrimidine dimers. In addition, a 3'-to-5' exonuclease purified from calf spleen greatly reduced synthesis on a UV-irradiated template in the presence of DNA polymerase alpha (176), suggesting that the mammalian exonuclease activity might act in a manner similar to that of the bacterial editing nuclease to remove bases inserted opposite dimers. This observation further suggests that this exonuclease activity might act *in vivo* to stall replication forks at sites of pyrimidine dimers in mammalian cells (see Section III,F). However, since the structure of the progeny molecules produced from UV-irradiated ϕX174 templates by mammalian cell polymerases was not examined (176), alternative explanations have not been eliminated. It is possible, for example, that the high level of DNA synthesis produced by the mammalian polymerases in the presence of UV damage might be due to initiation at sites other than that primed by the added restriction fragments, and that this type of synthesis might be inhibited by the addition of the spleen exonuclease.

Since the exiting exonucleases of prokaryotic DNA polymerases may control the replication of UV-irradiated DNA *in vitro*, we suggest that studies of polymerases carrying exonucleases with altered activities might substantially increase our understanding of this process. An ideal system for study might be the DNA polymerases from mutator and antimutator strains of bacteriophage T4. Polymerases from antimutator strains (*a*) have increased editing nuclease activity relative to polymerase activity; (*b*) replicate DNA more accurately; and (*c*) result in lower spontaneous mutation rates compared to normal polymerases (*179*). Conversely, polymerases from mutator strains have a reduced ratio of exonuclease to polymerase activity, replicate DNA less accurately, and produce higher spontaneous mutation rates *in vivo*. The effects of these altered exonuclease activities on replication of damaged templates have not been examined to date.

There is excellent recent evidence (*180*), employing DNA sequencing technology, that both prokaryotic and eukaryotic DNA polymerases stop DNA synthesis at sites of pyrimidine dimers during synthesis *in vitro*. In this analysis, single-stranded circular DNA of bacteriophage ϕX174 was primed with a restriction fragment of its complementary strand, and utilized for DNA synthesis by purified DNA polymerase I of *E. coli* or DNA polymerase alpha from human lymphoblastoid cells. The sequence of the molecules being synthesized was determined by the procedure of Sanger *et al.* (*181*). With UV-irradiated DNA templates, a new set of overlapping DNA fragments was obtained, each corresponding to a termination event occurring at the base preceding a tandem set of pyrimidines. These results clearly demonstrate that, in this purified *in vitro* system, both prokaryotic and eukaryotic DNA polymerases frequently stop DNA synthesis at the base preceding a pyrimidine dimer on the template strand, suggesting that an editing exonuclease may not be needed to prevent replication past sites of UV damage. These results are not in agreement with the kinetic analysis (*176*), possibly owing to differences in the enzyme sources or degree of purity, and in the procedures followed.

B. Methods for Measurement of Replication Fidelity

Identification of the mechanisms that control replication fidelity *in vivo* appears essential for an understanding of UV mutagenesis. The recent development of *in vitro* replication systems with the high fidelity observed *in vivo* (*182–184*) provide substantial information about spontaneous mutagenesis. These systems should be adaptable to studying fidelity *in vitro* with UV-irradiated templates.

In the past, replication fidelity was measured by the rate of incorporation of an incorrect nucleotide into a synthetic homopolymer. This type of assay of fidelity suffered from problems of high variability in the measured rate. The observed rates were several orders of magnitude higher than natural mutation rates and were extremely sensitive to the degree of purity of the deoxyribonucleoside triphosphates and the homopolymer template itself.

A recently developed *in vitro* system for measuring DNA replication fidelity avoids these difficulties by utilizing a natural template, the single-stranded DNA of bacteriophage ϕX174 (*182*). A sequenced amber mutant of this virus (*am*-3), defective in the virus lysis function, was primed with a restriction fragment whose 3'-hydroxyl end was 83 nucleotides from the site of the *am*-3 mutation, then used as a substrate for DNA synthesis by DNA polymerase I of *E. coli*. Insertion of an incorrectly paired base at the site of the amber mutation could revert the nonsense codon to either a missense codon or the original wild-type codon, thus producing an amino-acid substitution giving a functional protein. Such reversions were detected by transfecting cells with the newly replicated DNA. The number of revertants detected should reflect the fidelity of the replication system. It was found, first, that replication was highly accurate when the polymerase was presented with equimolar concentrations of the four dNTPs, and second, that increasing the concentration of an incorrect triphosphate promoted incorporation of an incorrect base. This result suggested that availability of each dNTP is an important factor in determining replication fidelity.

A difficulty with this system is that the level of insertion of incorrect bases may not be accurately measured by the number of revertant phage, because of mismatch repair in the infected cells removing the incorrectly inserted base (*176, 183*). This difficulty has been avoided by the development of an *in vitro* replication system employing the same amber mutant of ϕX174, but utilizing the double-stranded, replicative form of ϕX174 as a template and a replication complex composed of the gene products necessary for DNA synthesis in bacteriophage T4-infected cells (*183*). In this system, a rolling-circle type of replication produced long, single-stranded progeny DNA concatemers of several viral lengths. This single-stranded DNA was subsequently replicated in the *in vitro* system to form double-stranded progeny DNA. Individual molecules were cut from this replicated DNA by a restriction enzyme that makes one cut per viral genome. The complementary, "sticky" ends of these smaller molecules were annealed and ligated to form double-stranded circles, and reversion frequencies were determined by transfecting cells with them. This system may avoid possible biases due to mismatch repair in transfected cells, since mismatch repair is thought to discriminate between progeny and parental DNA strands on the basis of the higher extent of methylation of the latter (185). Consequently, a mismatch in a progeny strand should be removed more readily. The DNA produced in this assay is almost entirely composed of molecules in which both strands are progeny strands. Therefore, mismatch repair should act with equal probability on either strand.

The error frequency estimated from this assay (183) was very similar to the estimated *in vivo* fidelity, and may be attributed to the accessory proteins other than the T4 polymerase present in the polymerase complex used. The authors estimated the error frequency for an A · T to G · C transition to be less than 5 per 10^7 nucleotides copied, about 100-fold more accurate than that detected with the simpler assay (182) employing DNA polymerase I alone. This result strongly suggests that other replication proteins associated with DNA polymerase play an important role in the control of replication fidelity.

In a similar assay to study the replication of ϕX174 *am*-3 DNA, a reconstituted system of *E. coli* proteins necessary for DNA replication, consisting of DNA polymerase III, *rep* protein, and the single-stranded DNA binding protein of *E. coli*, was employed (184). This system generates single-stranded progeny molecules from a double-stranded, viral template. Cells were transfected with these molecules to determine the yield of revertants. This system was also as accurate as *in vivo* replication, and also depended upon the relative concentrations of added dNTPs. A kinetic analysis of the relationship between dNTP concentration and reversion frequency suggests that the reversion frequency depends on at least three factors: (*a*) incorrect insertion; (*b*) removal of the incorrect base by the editing exonuclease; and (*c*) polymerization of the next base downstream from the incorrectly inserted base, thus preventing nucleolytic removal of the mispaired base.

The *E. coli* single-strand-DNA binding-protein substantially enhances the fidelity of DNA synthesis by purified DNA polymerases with either synthetic polynucleotides or primed ϕX174 DNA as templates (186). This increased fidelity was observed with several purified DNA polymerases, including *E. coli* DNA polymerase I and calf thymus DNA polymerasas alpha and beta. Single-strand binding-protein binds cooperatively to DNA, removing secondary structure and enhancing synthesis. Evidently, these structural changes in the template also enhance fidelity. Dominant mutations in the gene coding for single-strand binding-protein (*lexC*) can affect UV sensitivity and UV-induced mutagenesis (*187*). One reasonable explanation for this result is that this protein might be modified or removed from DNA following irradiation (perhaps by the *recA* protein) to enhance replication of damaged DNA by relaxing replication fidelity.

With the availability of these unique systems for detecting replication fidelity, and knowledge of the entire DNA sequence of ϕX174, it should be possible to probe the behavior of DNA replication at sites of pyrimidine dimers in DNA and the mutagenic events that are likely to occur. It should also be possible to perform such analyses with purified eukaryotic DNA polymerases, although highly accurate, complex, *in vitro* systems for replicating simple viral molecules (such as that of SV40), similar to those just described for prokaryotic cells, have not yet been, but clearly should be, developed [see Note Added in Proof (c)].

VII. Summary and Future Perspectives

We have attempted to summarize the current literature on the topics of the replication and mutagenesis of UV-irradiated DNA in *E. coli* and in mammalian cells. Clearly, the mechanisms for these processes are much better understood in prokaryotic systems, largely owing to the availability of cellular mutants deficient in these processes and to the use of bacteriophages, which provide defined DNA templates for studying replication and replication fidelity. The recent use of animal viral systems to study the effects of UV irradiation on DNA synthesis and mutagenesis has already indicated how valuable these systems will be in elucidating replication and mutagenesis mechanisms in mammalian cells. Viral mutants, particularly in the larger viruses such as adeno- and herpes simplex virus, may also be useful for identifying viral functions that influence these processes. These functions may serve as models for the corresponding host-cell proteins.

We also wish to advocate the development and use of *in vitro* systems for analysis of DNA synthesis in mammalian cells. Since these systems should allow analysis of individual, purified components of the replication process, they should also be applicable to the study of these components in replication and mutagenesis of damaged DNA. Finally, as recombinant DNA technology continues to develop, it should be possible to clone and analyze the genes that are mutant in repair-deficient mammalian cells and to identify the gene products.

Acknowledgments

Research in the authors' laboratories is supported by National Institutes of Health Grants Nos. AG01689 and GM24496 and National Science Foundation Grant No. PCM 79-12059. The authors are grateful to Drs. P. Cerutti, J. Cleaver, H. Edenberg, Y. Fujiwara, P. Hanawalt, C. D. Lytle, R. Painter, A. Sarasin, R. Setlow, W. Summers, J. Trosko, and E. Waldstein for kindly sending us data prior to publication to be used in this review. We also wish to express our appreciation to Drs. J. Little and G. T. Bowden for critical reading of this manuscript and to I. Sassoon for assistance in its preparation.

References

1. L. E. Orgel, *Nature* **243**, 441 (1973).
2. J. B. Little, *Gerontology* **22**, 28 (1976).
3. P. C. Hanawalt, P. K. Cooper, A. K. Ganesan and C. A. Smith, *ARB* **48**, 783 (1979).
4. J. E. Cleaver, *BBA* **516**, 489 (1978).
5. B. M. Sutherland, *Int. Rev. Cytol.* Suppl. **8**, 301 (1978).
6. L. Grossman, A. Braun, R. Feldberg and I. Mahler, *ARB* **44**, 19 (1975).
7. T. Lindahl, This Series **22**, 135 (1979).
8. J. J. Roberts, *Adv. Radiat. Biol.* **7**, 211 (1978).
9. P. A. Swenson, *Photochem. Photobiol. Rev.* **1**, 269 (1979).
10. M. C. Paterson, *Adv. Radiat. Biol.* **7**, 1 (1978).
11. C. F. Arlett and A. R. Lehmann, *Annu. Rev. Genet.* **12**, 95 (1978).
12. R. B. Setlow, *Nature* **271**, 713 (1978).
13. R. F. Kimball, *Mutat. Res.* **55**, 85 (1978).
14. R. W. Hart, K. Y. Hall and F. B. Daniel, *Photochem. Photobiol.* **28**, 131 (1978).
15. M. W. Lieberman, *Int. Rev. Cytol.* **45**, 1 (1976).
16. K. C. Smith, *Photochem. Photobiol.* **28**, 121 (1978).
17. E. M. Witkin, *Bacteriol. Rev.* **40**, 869 (1976).
18. C. M. Radding, *ARB* **47**, 847 (1978).
19. C. Bernstein, *Microbiol. Rev.* (in press).
20. E. Moustacchi, R. Waters, M. Heude and R. Chanet, in "Radiation Research: Biomedical, Chemical and Physical Perspectives" (O. F. Nygaard, H. I. Adler, and W. K. Sinclair, eds.), p. 632. Academic Press, New York, 1976.
21. J. E. Cleaver, *Adv. Radiat. Biol.* **4**, 1 (1974).
22. R. B. Painter, *Genetics* **78**, 139 (1974).
23. R. R. Hewitt and R. E. Meyn, *Adv. Radiat. Biol.* **7**, 153 (1978).
24. R. Beukers and W. Berends, *BBA* **49**, 181 (1961).
25. E. Ben-Hur and R. Ben-Ishai, *BBA* **166**, 9 (1968).
27. E. A. Waldstein, S. Peller and R. B. Setlow, *PNAS* **76**, 3746 (1979).
28. E. C. Friedberg, K. H. Cook, J. Duncan and K. Mortelmans, *Photochem. Photobiol. Rev.* **2**, 263 (1977).
29. I. F. Ness and J. Nissen-Meyer, *BBA* **520**, 111 (1978).
29a. W. E. Cohn, N. J. Leonard, and S. Y. Wang, *Photochem. Photobiol.* **19**, 89 (1974).
30. R. B. Kelly, M. R. Atkinson, J. A. Huberman and A. Kornberg, *Nature* **224**, 495 (1969).
31. D. M. Livingston and C. C. Richardson, *JBC* **250**, 470 (1975).
32. J. W. Chase and C. C. Richardson, *JBC* **249**, 4553 (1974).
33. J. Doniger and L. Grossman, *JBC* **251**, 4579 (1976).
34. K. H. Cook and E. C. Friedberg, *Bchem.* **17**, 850 (1978).

35. A. Weissbach, *ARB* **46**, 25 (1977).
36. P. Howard-Flanders and R. P. Boyce, *Radiat. Res., Suppl.* **6**, 156 (1966).
37. W. D. Rupp and P. Howard-Flanders, *JMB* **31**, 291 (1968).
38. W. D. Rupp, C. E. Wilde, III, D. L. Reno and P. Howard-Flanders, *JMB* **61**, 25 (1971).
39. A. K. Ganesan, *JMB* **87**, 103 (1974).
40. A. K. Ganesan and P. C. Seawell, *Mol. Gen. Genet.* **141**, 189 (1975).
41. K. C. Smith and D. H. C. Meun, *JMB* **51**, 459 (1970).
42. R. H. Rothman and A. J. Clark, *Mol. Gen. Genet.* **155**, 279 (1977).
43. A. J. Clark, *Annu. Rev. Genet.* **7**, 67 (1973).
44. D. W. Mount, K. B. Low and S. H. Edmiston, *J. Bact.* **112**, 886 (1972).
45. L. J. Gudas and D. W. Mount, *PNAS* **74**, 5280 (1977).
46. C. A. Selsky and S. Greer, *Mutat. Res.* **50**, 395 (1978).
47. P. J. Abrahams and A. J. Van Der Eb, *Mutat. Res.* **35**, 13 (1976).
48. R. S. Day, *Nature* **253**, 748 (1975).
49. C. D. Lytle, S. A. Anderson and G. Harvey, *Int. J. Radiat. Biol.* **22**, 159 (1972).
50. K. H. Kraemer, E. A. DeWeerd-Kastelein, J. H. Robbins, W. Keijzer, S. F. Barrett, R. A. Petinga and D. Bootsma, *Mutat. Res.* **33**, 327 (1975).
51. S. Arase, T. Kozuka, K. Tanaka, M. Ikenaga and H. Takebe, *Mutat. Res.* **59**, 143 (1979).
52. W. Keijzer, N. G. J. Jaspers, P. J. Abrahams, A. M. R. Taylor, C. F. Arlett, B. Zelle, H. Takebe, P. D. S. Kinmont and D. Bootsma, *Mutat. Res.* **62**, 183 (1979).
53. A. J. Fornace, Jr., K. W. Kohn and H. E. Kann, Jr., *PNAS* **73**, 39 (1976).
54. R. S. Day, *Nature* **253**, 748 (1975).
55. C. F. Arlett, S. A. Harcourt and B. C. Broughton, *Mutat. Res.* **33**, 341 (1975).
56. A. R. Lehmann, S. Kirk-Bell, C. F. Arlett, M. C. Patterson, P. H. M. Lohman, E. A. de Weerd-Kastelein and D. Bootsma, *PNAS* **72**, 219 (1975).
57. A. R. Lehmann, S. Kirk-Bell, C. F. Arlett, S. A. Harcourt, E. A. de Weerd-Kastelein, W. Keijzer and P. Hall-Smith, *Cancer Res.* **37**, 904 (1977).
58. C. W. Dingman and T. Kakunaga, *Int. J. Radiat. Biol.* **30**, 55 (1976).
59. R. B. Setlow, P. A. Swenson and W. L. Carrier, *Science* **142**, 1464 (1963).
60. P. A. Swenson and R. B. Setlow, *JMB* **15**, 201 (1966).
61. K. G. Lark and C. A. Lark, *CSHSQB* **43**, 537 (1978).
62. K. C. Smith and D. H. C. Meun, *JMB* **51**, 459 (1970).
63. V. N. Iyer and W. D. Rupp, *BBA* **228**, 117 (1971).
64. P. Howard-Flanders, W. D. Rupp, B. M. Wilkins and R. S. Cole, *CSHSQB* **33**, 195 (1968).
65. M. Ohki and J. I. Tomizawa, *CSHSQB* **33**, 651 (1968).
66. W. D. Rupp and G. Ihler, *CSHSQB* **33**, 647 (1968).
67. B. M. Wilkins, S. E. Hollom and W. D. Rupp, *J. Bact.* **107**, 505 (1971).
68. B. M. Wilkins and P. Howard-Flanders, *Genetics* **60**, 243 (1968).
69. P. F. Lin and P. Howard-Flanders, *Mol. Gen. Genet.* **146**, 107 (1976).
70. R. G. Lloyd, *J. Bact.* **134**, 929 (1978).
71. P. Gaillet-Fauquet, M. Defais and M. Radman, *JMB* **117**, 95 (1977).
72. M. L. Gefter, *ARB* **44**, 45 (1975).
73. T. Kogoma, T. A. Torrey and M. J. Connaughton, *Mol. Gen. Genet.* **176**, 1 (1979).
74. G. M. Weinstock, K. McEntee and I. R. Lehman, *PNAS* **76**, 126 (1979).
75. K. McEntee, G. M. Weinstock and I. R. Lehman, *PNAS* **76**, 2615 (1979).
76. T. Shibata, C. DasGupta, R. P. Cunningham and C. M. Radding, *PNAS* **76**, 1638 (1979).

77. T. Shibata, R. P. Cunningham, C. DasGupta and C. M. Radding, *PNAS* **76**, 5100 (1979).
78. R. P. Cunningham, T. Shibata, C. DasGupta and C. M. Radding, *Nature* **281**, 191 (1979).
79. M. Volkert, A. J. Clark, and L. Margossian, *PNAS* (in press).
80. H. Ginsburg, S. Edmiston and D. Mount, *PNAS* (submitted for publication).
81. J. W. Roberts and C. W. Roberts, *PNAS* **72**, 147 (1975).
82. J. W. Roberts, C. W. Roberts and N. L. Craig, *PNAS* **75**, 4714 (1978).
83. J. W. Little, S. H. Edmiston, L. Z. Pacelli and D. W. Mount, *PNAS* **77**, 3225 (1980).
84. K. McEntee, *PNAS* **74**, 5275 (1977).
85. P. T. Emmerson and S. C. West, *Mol. Gen. Genet.* **155**, 77 (1977).
86. S. C. West, E. Cassuto, J. Mursalim and P. Howard-Flanders, *PNAS* (in press).
87. H. Potter and D. Dressler, *CSHSQB* **43**, 969 (1978).
88. R. Holliday, *Genet. Res.* **5**, 282 (1964).
89. R. D. Ley, *Photochem. Photobiol.* **18**, 87 (1973).
90. H. J. Edenberg and J. A. Huberman, *Annu. Rev. Genet.* **9**, 245 (1975).
91. R. A. Walters and C. E. Hildebrand, *BBRC* **65**, 265 (1975).
92. R. G. Martin and A. Oppenheim, *Cell* **11**, 859 (1977).
93. R. E. Meyn, R. R. Hewitt, L. F. Thomson and R. M. Humphrey, *Biophys. J.* **16**, 517 (1976).
94. A. M. Rauth, *Curr. Top. Radiat. Res.* **6**, 97 (1970).
95. J. Doniger, *JMB* **120**, 433 (1978).
96. S. D. Park and J. E. Cleaver, *NARes* **6**, 1151 (1979).
97. J. E. Cleaver, G. H. Thomas and S. D. Park, *BBA* **564**, 122 (1979).
98. D. Nathans and K. J. Danna, *Nature NB* **236**, 200 (1972).
99. A. R. Sarasin and P. C. Hanawalt, *JMB* **138**, 299 (1980).
99a. L. V. Crawford, N. C. Cole, A. E. Smith, E. Paucha, P. Tegtmeyer, K. Runcell and P. Berg, *PNAS* **75**, 117 (1978).
99b. C. J. Lai and D. Nathans, *CSHSQB* **39**, 53 (1974).
99c. C. J. Lai and D. Nathans, *Virology* **66**, 70 (1975).
100. L. F. Povirk and R. B. Painter, *Biophys. J.* **16**, 883 (1976).
101. H. J. Edenberg, *Biophys. J.* **16**, 849 (1976).
102. A. R. Lehmann, *JMB* **66**, 319 (1972).
103. S. N. Buhl, R. M. Stillman, R. B. Setlow and J. D. Regan, *Biophys. J.* **12**, 1183 (1972).
104. J. M. Clarkson and R. R. Hewitt, *Biophys. J.* **16**, 1155 (1976).
105. S. F. H. Chiu and A. M. Rauth, *BBA* **259**, 164 (1972).
106. G. T. Bowden, B. Giesselbach and N. E. Fusenig, *Cancer Res.* **38**, 2709 (1978).
107. S. D. Park and J. E. Cleaver, *PNAS* **76**, 3927 (1979).
108. J. I. Williams and J. E. Cleaver, *Mutat. Res.* **52**, 301 (1978).
109. J. E. Cleaver and G. H. Thomas, *BBRC* **36**, 203 (1969).
110. Y. Fujiwara, *Exp. Cell Res.* **75**, 483 (1972).
111. S. N. Buhl, R. B. Setlow and J. B. Regan, *Int. J. Radiat. Biol.* **22**, 417 (1972).
112. R. Meneghini, *BBA* **425**, 419 (1976).
113. R. Meneghini and P. Hanawalt, *BBA* **425**, 428 (1976).
114. R. Waters and J. D. Regan, *BBRC* **72**, 803 (1976).
115. Y. Fujiwara and M. Tatsumi, *Mutat. Res.* **43**, 279 (1977).
116. A. R. Lehmann and S. Kirk-Bell, *Photochem. Photobiol.* **27**, 297 (1978).
117. J. Rommelaere and A. Miller-Faures, *JMB* **98**, 195 (1975).
118. Y. Fujiwara and M. Tatsumi, *Mutat. Res.* **37**, 91 (1976).
119. U. B. Das Gupta and W. C. Summers, *Mol. Gen. Genet.* (in press).

120. J. D. Hall, J. Featherston and R. Almy, *Virology* **105**, 490 (1980).
121. D. R. Dubbs, M. Rachmeler and S. Kit, *Virology* **57**, 161 (1974).
122. H. Yamamoto and H. Shimojo, *Virology* **45**, 529 (1971).
123. C. A. Selsky, P. Henson, R. R. Weichselbaum and J. B. Little, *Cancer Res.* **39**, 3392 (1979).
124. P. A. Schaffer, V. C. Carter and M. C. Timbury, *J. Virol.* **58**, 490 (1978).
125. A. J. Levine, *BBA* **458**, 213 (1976).
127. E. Winocour, N. Frenkel, S. Ravi, M. Osenholts and S. Rozenblatt, *CSHSQB* **39**, 101 (1974).
128. N. P. Higgins, K. Kato and B. Strauss, *JMB* **101**, 417 (1976).
129. K. Tatsumi and B. Strauss, *NARes.* **5**, 331 (1978).
130. T. Nilsen and C. Baglioni, *JMB* **133**, 319 (1979).
131. S. M. D'Ambrosio and R. B. Setlow, *PNAS* **73**, 2396 (1976).
132. C. C. Chang, S. M. D'Ambrosio, R. Schultz, J. E. Trosko and R. B. Setlow, *Mutat. Res.* **52**, 231 (1978).
133. S. M. D'Ambrosio, P. M. Aebersold and R. B. Setlow, *Biophys. J.* **23**, 71 (1978).
134. R. B. Painter, *Nature* **275**, 243 (1978).
135. S. N. Buhl, R. B. Setlow and J. D. Regan, *Biophys. J.* **13**, 1265 (1973).
136. R. Lehmann and S. Kirk-Bell, *EJB* **31**, 438 (1972).
137. R. E. Meyn and R. M. Humphrey, *Biophys. J.* **11**, 295 (1971).
137a. E. M. Witkin, *Annu. Rev. Genet.* **3**, 525 (1969).
138. Z.-I. Horii and A. J. Clark, *JMB* **80**, 327 (1973).
139. L. Z. Pacelli, S. H. Edmiston and D. W. Mount, *J. Bact.* **137**, 568 (1979).
140. D. W. Mount, *PNAS* **74**, 300 (1977).
141. C. Coulondre and J. H. Miller, *JMB* **117**, 577 (1977).
142. Y. Ishii and S. Kondo, *Mutat. Res.* **27**, 27 (1975).
143. T. Kohno and J. R. Roth, *JMB* **89**, 17 (1974).
144. N. Kleckner, *Cell* **11**, 11 (1977).
145. N. Kleckner, K. Reichardt and D. Botstein, *JMB* **127**, 89 (1979).
146. A. W. Hsie, P. A. Brimer, T. J. Mitchell and D. G. Gosslee, *Somatic Cell Genet.* **1**, 383 (1975).
147. A. A. van Zeeland and J. W. I. M. Simons, *Mutat. Res.* **35**, 129 (1976).
148. V. M. Maher, D. J. Dornery, A. L. Mendrala, B. Konze-Thomas and J. J. McCormick, *Mutat. Res.* **62**, 311 (1979).
149. V. M. Maher and J. J. McCormick, in "Biology of Radiation Carcinogenesis" (J. M. Yuhas, R. W. Tennant, and J. D. Regan, eds.), p. 129. Raven, New York, 1976.
150. J. J. McCormick and V. M. Maher, in "DNA Repair Mechanisms" (P. C. Hanawalt, E. C. Friedberg, and C. F. Fox, eds.), p. 739. Academic Press, New York, 1978.
151. T. W. Glover, C. C. Chang, J. E. Trosko and S. S.-L. Li, *PNAS* **76**, 3982 (1979).
152. V. M. Maher, L. M. Ouellette, R. D. Curren and J. J. McCormick, *Nature* **261**, 593 (1976).
153. B. C. Myhr, D. Turnbull and J. A. DiPaolo, *Mutat. Res.* **62**, 341 (1979).
154. J. E. Cleaver, *Mutat. Res.* **52**, 247 (1978).
155. M. Blanco and R. Devoret, *Mutat. Res.* **17**, 293 (1973).
156. R. J. Huskey, *Science* **164**, 319 (1969).
157. J. George, R. Devoret and M. Radman, *PNAS* **71**, 144 (1974).
158. M. Castellazzi, J. George and G. Buttin, *Mol. Gen. Genet.* **119**, 139 (1972).
159. E. S. Tessman and T. Ozaki, *Virology* **12**, 431 (1960).
159a. C. K. DasGupta and R. K. Poddar, *Mol. Gen. Genet.* **139**, 77 (1975).
159b. T. Kato and Y. Shinoura, *Mol. Gen. Genet.* **156**, 121 (1977).

160. H. Ichikawa-Ryo and s. Kondo, *JMB* **97**, 77 (1975).
161. T. Kato, R. H. Rothman and A. J. Clark, *Genetics* **87**, 1 (1977).
161a. J. F. Bleichrodt and W. S. D. Verheij, *Mol. Gen. Genet.* **135**, 19 (1974).
162. R. S. K. Chaganti, S. Schonberg and J. German, *PNAS* **71**, 4508 (1974).
163. C. D. Lytle, J. Coppey and W. D. Taylor, *Nature* **272**, 60 (1978).
164. W. P. Jeeves and A. J. Rainbow, *Mutat. Res.* **60**, 33 (1979).
165. L. E. Bockstahler and C. D. Lytle, *Photobiochem. Photobiol.* **25**, 477 (1977).
166. A. R. Sarasin and P. C. Hanawalt, *PNAS* **75**, 346 (1978).
167. U. B. DasGupta and W. C. Summers, *PNAS* **75**, 2378 (1978).
168. C. D. Lytle and J. G. Goddard, *Photochem. Photobiol.* **29**, 959 (1979).
169. A. Sarasin and A. Benoit, *Mutat. Res* **70**, 71 (1980).
170. R. Risser and C. Mulder, *Virology* **58**, 424 (1974).
171. W. W. Brockman and D. Nathans, *PNAS* **71**, 942 (1974).
172. G. Khoury, G. C. Fareed, K. Berry, M. A. Martin, T. N. H. Lee and D. Nathans, *JMB* **87**, 289 (1974).
173. W. Fiers, R. Contreras, G. Haegeman, R. Rogiers, A. Van de Voorde, H. Van Heuverswyn, J. Van Herreweghe, G. Volckaert and M. Ysebaert, *Nature* **273**, 113 (1978).
174. C. D. Lytle, J. G. Goddard and C.-H. Lin, *Mutat. Res.* **70**, 139 (1980).
175. M. Wagner, J. Skare and W. C. Summers, *CSHSQB* **39**, 683 (1974).
176. M. Radman, G. Villani, S. Boiteux, A. R. Kinsella, B. W. Glickman and S. Spadari, *CSHSQB* **43**, 937 (1978).
177. M. G. Sarngadharan, M. Robert-Guroff and R. C. Gallo, *BBA* **516**, 419.
178. D. Brutlag and A. Kornberg, *JBC* **247**, 241 (1972).
179. N. Muzyczka, R. L. Poland and M. J. Bessman, *JBC* **247**, 7116 (1972).
180. P. Moore and B. S. Strauss, *Nature* **278**, 664 (1979).
181. F. Sander, S. Nicklen and A. R. Coulson, *PNAS* **74**, 5463 (1977).
182. L. A. Weymouth and L. A. Loeb, *PNAS* **75**, 1924 (1978).
183. C. C. Liu, R. L. Burke, U. Hibner, J. Barry and B. Alberts, *CSHSQB* **43**, 469 (1978).
184. A. R. Fersht, *PNAS* **76**, 4946 (1979).
185. B. W. Glickman and M. Radman, *PNAS* **77**, 1063 (1980).
186. T. A. Kunkel, R. R. Meyer and L. A. Loeb, *PNAS* **76**, 6331 (1979).
187. B. F. Johnson, *Mol. Gen. Genet.* **157**, 91 (1977).

NOTE ADDED IN PROOF. (a) Recent evidence indicates that while the UV-specific endonuclease from *E. coli* exhibits the activity described, the incision step in *M. luteus* [W. A. Haseltine, L. K. Gordon, C. P. Lindan, R. H. Grafstrom, N. L. Shaper and L. Grossman, *Nature (London)* **285**, 634 (1980)] and in T4 bacteriophage-infected cells [P. C. Seawell, C. A. Smith and A. K. Ganesan, *J. Virol.* **35**, 790 (1980)] involves cleavage of (1) the glycosylic bond of the 5' pyrimidine nucleoside of the dimer and (2) the phosphodiester bond between the two pyrimidine nucleosides of the dimer.

(b) The *lexA* protein has recently been purified and has been shown by a variety of biochemical tests to act as a repressor of the *recA* gene (D. Mount and J. Little, personal communication).

(c) The recently developed *in vitro* replication system for human adenovirus may prove useful for analyzing replication of UV-damaged viral DNA [M.D. Challberg and T. J. Kelly, *JMB* **135**, 999 (1979); L. M. Kaplan, H. Ariga, J. Hurwitz and M. S. Horwitz, *PNAS* **76**, 5534 (1979)].

The Regulation of Initiation of Mammalian Protein Synthesis

ROSEMARY JAGUS
W. FRENCH ANDERSON AND
BRIAN SAFER

Laboratory of Molecular
Hematology,
National Heart, Lung, and Blood
Institute,
National Institutes of Health,
Bethesda, Maryland

I. Importance of Initiation in the Regulation of Protein Synthesis in Mammalian Tissues	129
II. Sequence of Events	133
A. Introduction	133
B. Production of Native 43 S Ribosomal Subunits (43 S_N)	135
C. Met-tRNA$_f$ Binding to 43 S_N Ribosomal Subunits	137
D. Binding of mRNA to 43 S Preinitiation Complex	142
E. 60 S Ribosomal Subunit Joining and Initiation Factor Release	151
III. Regulation of Initiation	153
A. Regulation of Globin mRNA Translation by Hemin	153
1. Historical Introduction	153
2. Reticulocyte Lysate	154
3. Inhibitor Formation	158
4. Phosphorylation of eIF-2α: Role of eIF-2 Phosphatase Activity	161
5. Relationship between eIF-2α Phosphorylation and Protein Synthesis	163
6. Reversal Factors	167
7. Summary of Mechanism of Hemin Control of Translation	167
B. Regulation of Translation by Interferon	169
1. Historical Introduction	169
2. Double-Stranded RNA-Dependent Protein Kinase(s)	170
3. (2'-5')-Oligoadenylate and (2'-5')-Oligoadenylate Synthetase	171
4. (2'-5')-Oligoadenylate-Activated Endonuclease Activity	173
5. (2'-5')-Phosphodiesterase Activity	173
6. Relative Importance of Double-Stranded RNA-Dependent Enzyme Activity	173
IV. Summary: Overview on Present Understanding of the Control of Initiation	175
References	177

Introduction[1]

Protein synthesis can be divided into three processes: (a) *initiation*, the attachment of ribosomal subunits and other necessary components to the 5' end of the coding sequence of mRNA, encompassing a series of events that ensure the recognition of the initiator AUG codon of mRNA and the decoding of the template in the correct reading frame; (b) *elongation*, the stepwise addition of amino acids, in a sequence determined by the base sequence of mRNA, to produce a growing polypeptide chain; (c) *termination*, the final stage, in which the ribosome and the newly synthesized polypeptide are released from the mRNA.

Because of the relative stability of mammalian ribosome and mRNA concentrations, rapid fluctuations in protein synthesis are achieved not by changes in the levels of these translational components, but by their activity. Kinetic analyses of mammalian protein synthesis have demonstrated that initiation is the rate-limiting step under steady-state conditions. In addition, changes in the rate of initiation affect the relative rates of translation of different mRNA species.

---

[1] Abbreviations:
MEL: mouse erythroleukemia
mRNP: messenger ribonucleoprotein particle
phytag: phytohemagglutinin
eIF: eukaryotic initiation factor (1, 1a)
$m^7G(5')ppp(5')Nm-N'$: general structure of methylated 5' terminus of eukaryotic mRNA ("cap")
$(2'-5')A_n$: $(2'-5')$-oligoadenylate
$tRNA_f$: the eukaryotic tRNA that accepts methionine to become the specific initiator tRNA, Met-$tRNA_f$
Met-$tRNA_f$: the specific eukaryotic initiator tRNA, which can be N-formylated by an enzyme from *E. coli*, although such a modification does not occur in eukaryotes *in vivo*
40 S ribosomal subunit: small ribosomal subunit derived by high-salt wash of polysomes; does not contain any initiation factors
43 S_N ribosomal subunits: small ribosomal subunit, competent to take part in initiation; contains eIF-3, eIF-4C, and Met-$tRNA_f$ hydrolase
43 S preinitiation complex: first preinitiation complex; contains (eIF-2) · GTP, eIF-3, eIF-4C, Met-$tRNA_f$, and Met-$tRNA_f$ hydrolase
48 S preinitiation complex: in addition to components found in 43 S preinitiation complex, contains mRNA and a variety of mRNP proteins
74 S ribosomal couple: represents pairs of 40 S and 60 S ribosomal subunits not holding mRNA, Met-$tRNA_f$, or initiation factors, and not participating in protein synthesis
80 S initiation complex: fully active complex (of 48 S and 60 S ribosomes) containing Met-$tRNA_f$, Met-$tRNA_f$ hydrolase and mRNA

These considerations have focused attention on the importance of initiation as a control point in protein synthesis. This chapter reviews current knowledge of the initiation of protein synthesis up to December 1979.

I. Importance of Initiation in the Regulation of Protein Synthesis in Mammalian Tissues

The rate of protein synthesis in animal cells or tissues is affected by a variety of agents and metabolic conditions. These include the age and sex of the animal, its nutritional and hormonal status, its state of health, and response to injury and environmental trauma. In addition, qualitative regulation, resulting in alterations in the relative amounts of different proteins synthesized, occurs during hormonal induction of enzymes and changes in the patterns of gene expression that accompany development and differentiation.

Generally, quantitative control of protein synthesis in a steady-state situation is regulated by the number of ribosomes per cell; and qualitative control, by the availability of mRNAs. However, during transition periods, when a cell or tissue is responding to a stimulus by moving from one steady-state condition to another, initial alterations in the rate of protein synthesis, by changes in the utilization of existing translational components, are of great importance; a rapid response is required, and alterations in protein synthetic rate are achieved by changes in the rate of initiation. Changes in the rate of protein synthesis initiation can also modulate the spectrum of proteins synthesized by changing the relative efficiency with which different mRNAs are translated (*1b*–*3*). In addition, changes in protein synthetic rate, induced by alterations in initiation rate, may ultimately influence the number of ribosomes and, therefore, the rate of protein synthesis in the new steady-state conditions (*4*–*7*).

Cellular ribosome content sets an upper limit on the rate of protein synthesis, and protein synthesis per ribosome as a function of time is relatively constant in tissues in a steady-state condition. For example, although the rate of protein synthesis varies sixfold in muscles of fed rats of various ages, the rate of synthesis per ribosome changes less than twofold (*8*). Also, cellular ribosomal content can be modulated, and, for long-term adaptations in the rates of protein synthesis, this appears to be the major regulatory mechanism (*9*, *10*). For instance, in animals growing at different rates because of nutritional limitations, the ribosome content of the tissue varies with the protein synthetic rate (*11*, *12*). Similarly, in the heart undergoing work hypertrophy, the ribosome content and protein synthetic rate rise during the period of growth and return to a lower level after the required muscle growth is achieved (13–16).

In animals transferred to a protein- or energy-deficient diet, ribosome content falls initially, then stabilizes at a level compatible with the new protein synthetic rate (9). However, the rate of protein synthesis in tissues from fasting animals decreases much more rapidly than the fall in ribosome content. This effect arises from a decrease in the rate of initiation of protein synthesis, made apparent by increases in the ratio of monosomes to polysomes, observed in both liver and skeletal muscle of fasted animals (8, 10, 11, 15, 17–20). Similarly, in liver after partial hepatectomy (21) or in kidney after unilateral nephrectomy (22), there is a rapid increase in the rate of protein synthesis, which is achieved at the level of initiation (21, 22).

There are many other conditions in which changes in the rate of protein synthesis initiation occur as part of the early response to a stimulus: many tissues respond to hypoxia and ischemia by a reversible reduction of initiation (23, 24); diabetes is similar to starvation in decreasing initiation in rat liver, heart, and skeletal muscle (25–30), an effect reversed by the administration of insulin (26–30); growth hormone also stimulates initiation in liver (31) and heart muscle (32).

Technical considerations make animal tissues inconvenient systems in which to elucidate the mechanisms underlying the control of initiation in mammalian cells; however, a variety of animal cell-culture systems are available in which rates of protein synthesis can be manipulated. For instance, in mammalian fibroblasts, grown in culture, protein synthesis can be stimulated by serum (33–35) or inhibited by amino-acid (36–40) or glucose starvation (36, 37, 41). Protein synthesis can be stimulated in T lymphocytes, in culture, by the plant lectin, phytohemagglutinin (phytag), a response considered equivalent to the antigenic stimulation of these cells (42).

However, some uncertainties arise in the use of the cultured cell systems as models for the study of regulation. For instance, serum may contain fetal growth stimulants not normally found in the adult, and also platelet-derived growth factors, presumably found in tissues only during the response to local injury (43–45). Similarly, amino-acid starvation of cultured cells does not provide an equivalent signal to that received by tissues of starved animals, since the individual tissues of a fasting animal are probably not deprived of nutrients; plasma glucose levels, for instance, are maintained by elaborate homeostatic mechanisms. Plasma and tissue amino-acid levels are also relatively well maintained during fasting (46, 47). It is more likely that tissues in fasting animals do not respond primarily to direct nutritional deprivation, but to an undefined hormonal mechanism that functions to protect the whole organism (48). However, the response of protein synthesis *in vitro* resembles the responses found in tissues stimulated *in vivo* sufficiently to provide technically simpler situations for detailed study of the mechanisms controlling protein synthesis. The characteristics of several such systems are summarized in the following section.

A. Amino-Acid Deprivation

In ascites (36, 37), HeLa (38), sarcoma (39), and Chang cells (40), deprivation of an essential amino acid inhibits protein synthesis and is characterized by increases in the percentage of inactive ribosomes. The inhibition of initiation is perceptible within 15 minutes (37). The rate of peptide chain elongation is somewhat reduced, but insufficient to account for the observed decreases in protein synthesis (38). Upon restoration of the essential amino acid, re-formation of polysomes is discernible within 2 minutes and approaches completion within 10–15 minutes (37–39, 49). The inhibition of initiation and its reversal proceed in the presence of inhibitors of RNA synthesis (36, 38–40), which suggests that the mechanism of inibition involves modification of preexisting translational components. Messenger RNA, released as mRNP during inhibition, can be used to re-form polysomes upon readdition of the amino acid (39, 49). After prolonged amino-acid deprivation, this reversal requires progressively more time, and is less complete in the presence of inhibitors of RNA synthesis (36, 40). This indicates that inhibition of the production of translational components becomes more important during long-term amino-acid starvation.

B. Glucose Deprivation

An increase in the ratio of monosomes to polysomes, indicative of an inhibition of initiation, occurs also in cultured cells deprived of glucose; these include Ehrlich ascites cells (36, 37), 3T3 cells (10), and thymic lymphocytes (41). The manifestation of inhibition requires 2–4 hours in Ehrlich ascites cells, presumably because of the availability of other metabolic substrates. Reversal of the inhibition occurs within minutes of the readdition of glucose (37). If glucose deprivation is prolonged, a reduction in ribosome content develops (36, 37).

C. Effects of Cell Density and Serum

During the stationary growth phase, protein synthesis in cultured fibroblasts decreases to a value between 25% and 30% of the rate observed in the exponential growth phase (50–53). Although this decrease can be partially accounted for by a lower ribosome content, the frequency of initiation is also very low (54). When the cells are stimulated with serum, increases in protein synthesis can be detected within an hour (34, 35, 54). Although ribosome content increases twofold over a 20–30-hour period, the early increases in protein synthesis are largely attributable to increases in the rate of initiation (55).

D. Effects of Phytohemagglutinin in T Lymphocytes

T lymphocytes are nondividing cells that remain in a nonproliferating state in culture unless induced to divide by phytohemagglutinin (phytag) (42). Although the ribosome content rises only one-half to twofold during the first 20 hours after addition of this inducer, the rate of protein synthesis increases seven- to tenfold, partially owing to the mobilization of the relatively large pool of inactive ribosomes in resting lymphocytes, and represents initiation (56–59). During the first 12 hours, when changes in mRNA content are not detectable, such increased initiation rates reflect increased initiation factor activity (56; reviewed in 60).

Major regulatory features are shared by these cell-culture systems and the tissue responses described above. These include (a) adjustment of the ribosome content to prolonged stimulus; and (b) initial regulation of the rate of protein synthesis during the transitory period of adjustment by change in the rate of initiation. In all these systems, the initial rapid changes occur independently of a change in mRNA content (50, 55, 56, 61).

Despite the valuable insights yielded by these systems, progress on the detailed mechanism and control of the initiation of protein synthesis has come largely from investigations utilizing a cell-free protein-synthesizing system derived from rabbit reticulocytes. This system offers several major advantages over the systems already described: reticulocytes are enucleated cells that are readily lysed to yield cell-free extracts containing little nuclease or protease activity; they provide the major source of material for the purification of mammalian initiation factors and other translational components; reticulocyte lysates synthesize protein at rates approaching those found in the intact cell, in contrast to those produced from nucleated cells; the protein synthetic rate in reticulocyte lysates is regulated by hemin. In addition, use of this cell-free system avoids the complications resulting from permeability barriers and compartmentalization in intact cells.

The data available from cell-culture systems are integrated with the current understanding of mechanism and control of protein synthesis derived from reticulocyte lysates in the concluding Section IV. Before discussing the regulation of initiation, the components and sequence of events involved in this process are presented.

II. Sequence of Events

A. Introduction

The sequence of events comprising the initiation of protein synthesis in mammalian systems can be reduced to four major steps, depicted schematically in Fig. 1 (62). First, a native 43 S_N ribosomal subunit is generated by the interaction of specific peptides with a 40 S ribosomal subunit. The initiator aminoacyl-tRNA species, Met-tRNA$_f$, then binds to a native ribosomal subunit to form the first preinitiation complex. This is followed by mRNA binding to form an intermediate preinitiation complex, which sediments at 48 S. The final step is the joining of a 60 S ribosomal subunit to complete the formation of an 80 S initiation complex.

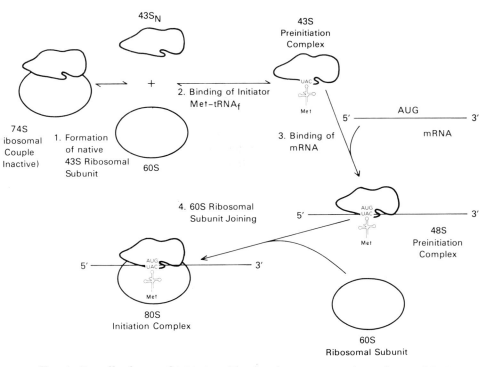

FIG. 1. Overall scheme of initiation. The four basic steps in the pathway of 80 S initiation complex are shown. The small ribosomal subunit is depicted according to the model of Emanilov et al. (62). Figures 2, 3, 4, and 6 illustrate each step in detail, showing initiation factor involvement.

TABLE I
Eukaryotic Initiation Factors

Factor		Molecular weight		
Nomenclature[a]	Other names[b]	Dodecyl sulfate	Native	Function (where known)
eIF-1	IF-E1	15,000	15,000	Stabilization of 48 S preinitiation complex
eIF-2	IF-MP, IFL3, IF1, IF-1, IF-E2, EIF3, EIF2	α 32,000 β 35,000 γ 55,000	125,000	Activated by GTP; necessary for Met-tRNA$_f$ binding to 43 S_N subunits
eIF-2A	—	65,000	65,000	Not needed for translation of natural mRNA; precise physiological role uncertain
eIF-3	IF-MS, IF-E3, IF-3, EIF-3	35,000 39,000 43,000 49,000 69,000 110,000 130,000	700,000	Formation of 43 S_N. Implicated in mRNA binding to 43 S preinitiation complex, but this property could be due to contaminating factor(s)
eIF-4A	IF-EMC, IF-M4, IF-E4	50,000	50,000	mRNA binding
eIF-4B	IF-M3, IF-E6	80,000	80,000	mRNA binding by cap recognition (this factor may contain two different activities)
eIF-4C	IF-M2Bβ, IF-E7	19,000	17,000	Formation of 43 S_N. Some involvement in stabilization of 48 S preinitiation complex. Stimulates subunit joining
eIF-4D	—	17,000	15,000	Not required for translation of natural mRNA; physiological role uncertain
eIF-4E (postulated)[c]	—	24,000	24,000	Minor contaminant of eIF-3 and eIF-4B, cap-binding protein
eIF-5	IF-M2A, IF-L2, IF-E5, IF-3, IF-11, F-0.25, IF-S2	150,000	125,000	Initiation factor release, 60 S subunit joining

[a] Adopted at International Symposium on Protein Synthesis, 1976. (See also *1a*.)
[b] Under which factor has been described.
[c] Contaminant of eIF-3 and eIF-4B; may be a new factor.

Assembly of the 80 S initiation complex is catalyzed by specific initiation factors. Purification of the factors required for the 80 S initiation complex assembly has been carried out in four laboratories from both rabbit reticulocyte lysate (63–79) and Krebs ascites cells (80). Nine protein factors have been described and classified so far.[2] Five of these factors are required (eIF-2, -3, -4A, -4B, -5); two have only stimulatory effects (eIF-1 and eIF-4C). The remaining two (eIF-2A and eIF-4D) are active only in model assay systems with artificial templates; their physiological function is unclear. The physical and biochemical properties of these nine factors, summarized in Table I, is presented in detail in Section II. A polypeptide of molecular weight 24,000, a trace contaminant of eIF-3 and 4B that cross-links specifically to capped viral mRNA has recently been found (81–84). This is tentatively classified in Table I as eIF-4E.

In addition to ribosomal subunits, Met-tRNA$_f$, mRNA, and the initiation factors, GTP and ATP are also required for the initiation of protein synthesis.

B. Production of Native 43 S Ribosomal Subunits (43 S$_N$)

The first step in the formation of the 80 S initiation complex is the generation of native, small ribosomal subunits, illustrated in Fig. 2. During polypeptide chain termination, 80 S ribosomes tran-

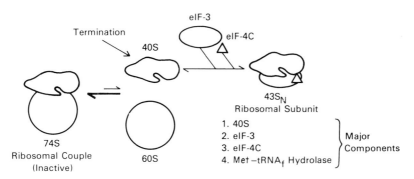

FIG. 2. Formation of 43 S$_N$ ribosomal subunit. This depicts the involvement of eIF-3, eIF-4C, and Met-tRNA$_f$ in the first step in 80 S initiation complex formation, the production of a small ribosomal subunit competent to take part in initiation. Binding of eIF-3 and eIF-4C to form 43 S$_N$ prevents the spontaneous reassociation of 40 S and 60 S ribosomal subunits, which occurs under *in vivo* ionic conditions.

---

[2] A uniform nomenclature for these factors, used throughout this review, was proposed at an international symposium on protein synthesis in 1976 (1) and is in general use (1a). See Table I.

siently dissociate into free subunits upon release from mRNA (85). Mammalian ribosomal subunits possess a high affinity for each other *in vivo*, with the equilibrium favoring the formation of nonfunctional 74 S ribosomal couples (86), distinguishable from functional 80 S monoribosomes by sucrose density gradient centrifugation (87–89). Dissociation of 74 S nonfunctional couples to 40 S and 60 S subunits by nonribosomal proteins is an obligatory first step for all subsequent stages of initiation complex formation (85, 86, 90–93).

The specific initiation factor responsible for formation of active 43 S_N ribosomal subunits appears to be eIF-3, a large polypeptide complex, isolated and referred to previously as IF-E_3 (70, 71, 75), IF-M5 (64), IF-3 (94), and EIF-3 (95).[2] Characterization of this factor from rabbit reticulocytes indicates that it is homogeneous as judged by both electrophoresis under nondissociating conditions and sedimentation analysis (96, 97). Native eIF-3 has a molecular weight of approximately 700,000, corresponding to a sedimentation coefficient of 15–17 S (64, 70, 73, 94). Electrophoresis in the presence of dodecyl sulfate reveals nine major polypeptide components ranging in mass from 35,000 to 160,000 daltons (73, 96, 97). The stoichiometry of the peptides indicates the presence of one copy per molecule of five of the individual peptides, two copies of two, and less than one each of the remaining two (73). Three of the peptides may be phosphorylated (74).

Evidence for a role of eIF-3 as an anti-association factor in the formation of 43 S_N ribosomal subunits, rather than as a dissociation factor, was suggested by kinetic experiments demonstrating that a partially purified protein fraction containing eIF-3 is much more efficient in recycling ribosomal subunits through polysomes than single ribosomes (98). Anti-association is also suggested by the stoichiometric association of one eIF-3 complex per 43 S_N ribosomal unit. This has been established by electrophoresis of 43 S_N ribosomal subunits (88, 96, 97, 99), use of radiolabeled factor (97), and electron microscopy (62). It is also suggested by the difference in buoyant density between salt-washed ribosomes (1.51) and 43 S_N ribosomal subunits (1.40–1.42), which can be calculated to be due to an extra 900,000-dalton protein, a value close to the mass of eIF-3.

The exact mechanism by which eIF-3 interacts with 40 S ribosomal subunits is not known, although it does so independently of ATP or GTP and does not require other initiation factors (70, 73, 96, 97). At least eight, and probably all nine, of the major polypeptides of eIF-3 are found in 43 S_N ribosomal subunits (96, 97). Among current questions are why the factor is so large, and whether different functional roles exist for the individual polypeptide components. However, electron-microscope observations of 43 S_N ribosomal subunits indicate that eIF-3 binds to the 40 S subunit in the region that interfaces with the 60 S ribosomal subunit in a functional ribosome (62). The binding of eIF-3 may sterically inhibit nonspecific ribosomal subunit association. Since the interfacing region is considered to be a functional site in the initiation process, as well as the region of mRNA localization during translation (62), it is possible that eIF-3 serves to organize the interaction of other initiation factors at the active site of translation.

Another factor, eIF-4C (see Table I), appears to be involved in the production of 43 S_N ribosomal subunits in conjunction with eIF-3 (100). Also designated IF-12Bβ (67), eIF-4C is a small protein of mass 17,000–19,000 daltons (67, 77). It is reported to be an acidic protein with an isoelectric point of 5.6 (77), although it behaves like a basic protein on phosphocellulose (97). In the presence of eIF-3, it stimulates the dissociation of 74 S ribosomal couples and can be found in amounts equivalent to eIF-2, eIF-3, and Met-tRNA$_f$ in 43 S preinitiation complexes (100).

The entire initiation sequence can occur in the absence of eIF-4C, but the efficiency of the overall reaction is considerably lower (100). The increased production of 43 S_N ribosomal subunits is not sufficient to account for the observed differences in 80 S initiation complex formation in the presence of eIF-4C. In addition, this factor stabilizes 43 S preinitiation complexes, giving these complexes a selective advantage in all partial initiation reactions tested (100) (see Sections II,C,3; II,E). The presence of a Met-tRNA$_f$ deacylase, bound to 43 S_N ribosomal subunits, has been reported by several laboratories (101–104). Reports of the mass of the native enzyme vary from 80,000 to 170,000 daltons (105, 106). Although its location on 43 S_N ribosomal subunits suggests a role in protein synthesis, its relationship to the events of initiation are not understood. However, it is known that under physiological conditions, Met-tRNA$_f$ is not accessible to the deacylase in the ternary or 43 S preinitiation complexes (105, 106). Additionally, Met-tRNA$_f$ is not accessible to the deacylase in the 80 S initiation complex when the complex is formed with natural mRNA (106).

C. Met-tRNA$_f$ Binding to 43 S_N Ribosomal Subunits

1. Properties of eIF-2

The second step of initiation is the binding of the specific initiator tRNA (Met-tRNA$_f$) to 43 S_N ribosomal subunits as depicted in Fig. 3.

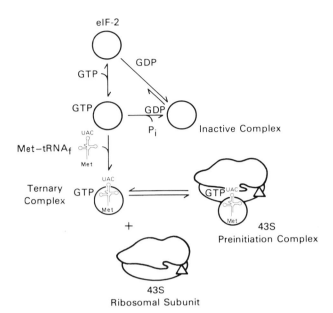

FIG. 3. Formation of 43 S_N preinitiation complex. This depicts the second step in initiation, the binding of the initiator tRNA, Met-tRNA$_f$, to the small ribosomal subunit. Before interaction with 43 S_N ribosomal subunit, Met-tRNA$_f$ forms a stable ternary complex with eIF-2 and GTP. GDP, which has a higher affinity for eIF-2 than GTP, is a potent inhibitor of ternary complex formation.

This is accomplished in distinct stages by the intervention of the initiation factor eIF-2. There have been many reports of the isolation of eIF-2, which has been designated IF-MP (64, 107), IFL3 (108), IF1 (109), IF-1 (110), IF-E2 (70), EIF3 (111), and EIF2 (112).[2] Homogeneous eIF-2 has a native molecular weight of 122,000 as determined by chemical cross-linking and electrophoresis, amino-acid analysis and reconstitution of model molecules, and sedimentation equilibrium centrifugation (113). The conclusion from these studies is that the biological form of eIF-2 is an α, β, γ oligomer with a molecular weight of 122,000 (113). By sedimentation equilibrium analysis, the molecular weights of the subunits are: α, 32,000; β, 35,000; γ, 55,000, as illustrated in Table II (113). This compares with previously reported values of α, 32,000–37,000; β, 49,000–59,000; γ, 52,000–55,000 (64,70,75,79) (see Table I).

The molecular weights obtained by sedimentation equilibrium centrifugation for all subunits are in reasonable agreement with those determined by electrophoresis in acid urea and with those of α and γ

TABLE II
CHARACTERISTICS OF EUKARYOTIC INITIATION FACTOR 2 (eIF-2)[a]

Characteristic	eIF-2 Subunits			Native eIF-2
	α	β	γ	$\alpha\beta\gamma$
Molecular weight (M_r)	32,000	35,000	55,000	122,000
pI	5.1	5.4	8.9	6.4
Gel system	Apparent M_r by polyacrylamide gel electrophoresis			
a. pH 3.5, 6 M urea	36,000	36,000	55,000	
b. pH 7, dodecyl sulfate, 50 mM sodium phosphate	35,000	55,000	55,000	
c. pH 7, dodecyl sulfate, 100 mM sodium phosphate	32,000	49,000	55,000	
d. pH 8.3, dodecyl sulfate, Tris/glycine	35,000	59,000	52,000	

[a] Data from Lloyd et al. (113).

in dodecyl sulfate. However, the estimated molecular weight of the β subunit in dodecyl sulfate at neutral or alkaline pH ranges from 49,000 to 59,000. The resolution of eIF-2 subunits by electrophoresis under various conditions is shown in Table II. Because of the anomalous behavior of the β subunit in dodecyl sulfate gels, which may be related to the elevated content of basic amino acids (20%), the most reliable characteristic for distinguishing them is the isoelectric point: α, 5.1; β, 5.4; γ, 8.9; $\alpha\beta\gamma$, 6.4 (113).

Isolation of individual subunits by isoelectric focusing has suggested that the α subunit binds guanine nucleotides, while the β subunit possesses Met-tRNA$_f$ binding activity (114). The function of the γ subunit is not known. However, dissociation of the subunits results in a dramatic loss of Met-tRNA$_f$ binding activity ($>90\%$) and an inability to reconstitute the native oligomers (114). This has precluded an unambiguous demonstration that all three subunits are necessary for Met-tRNA$_f$ binding to 43 S_N ribosomal subunits, although all three subunits are found in 43 S preinitiation complexes (96, 99). Reports have appeared on the isolation of eIF-2 containing two subunits only, which seem to be completely active (115).

2. TERNARY COMPLEX FORMATION

Before interaction with 43 S_N ribosomal subunits, Met-tRNA$_f$ forms a stable ternary complex with eIF-2 and GTP (107, 109, 116,

117). Binding proceeds as an ordered sequence, during which GTP binding precedes that of Met-tRNA$_f$ (*107*). A binary complex between eIF-2 and Met-tRNA$_f$ can be formed *in vitro* that will subsequently bind to 43 S$_N$ ribosomal subunits (*107*), although GTP increases the rate of Met-tRNA$_f$ binding to 43 S$_N$ ribosomal subunits 20-fold (*107*). Nonhydrolyzable GTP analogs may substitute for GTP in both ternary complex formation and its binding to 43 S$_N$ ribosomal subunits (*75, 107*).

GDP is a potent inhibitor of ternary complex formation ($K_i = 3.4 \times 10^{-7}$ M) and a competitive inhibitor of GTP binding (*118–121*). eIF-2 has a higher affinity for GDP ($K_D\text{GDP} = 3.1 \times 10^{-8}$ M) than GTP ($K_D\text{GTP} = 2.5 \times 10^{-6}$ M) (*118, 119*). Consequently, *in vivo*, the rate of initiation may be regulated by GTP/GDP ratios and, thus, ultimately, the energy status of the cell. However, although studies with reconstituted, fractionated systems suggest that initiation is more sensitive to changes in guanylate energy charge ([GTP]/[GTP][GDP]) than elongation, investigation of rabbit reticulocyte lysate indicates that elongation is the more sensitive step (*121*).

The ternary complex, (eIF-2) · GTP · (Met-tRNA$_f$), is stable and binds to nitrocellulose filters. Inclusion of the radiolabeled Met-tRNA$_f$ in the reaction is the basis for the ternary complex assay used in many laboratories (*96, 97, 107, 122, 123*). However, caution is necessary in the interpretation of the results of such studies. This assay is used to assess the extent of ternary complex formation only; i.e., it is used to determine what percentage of eIF-2 is active under a particular set of experimental conditions. The eIF-2 preparations isolated in numerous laboratories are not usually fully active in ternary complex formation, as measured by the retention of radiolabeled Met-tRNA$_f$ on nitrocellulose filters; values of 5% to 70% have been reported (*96, 97, 107, 117, 122, 123*).

Low eIF-2 efficiency in ternary complex formation has led to the isolation of putative stimulatory cofactors that appear to increase the extent of Met-tRNA$_f$ binding (*124–132*).[2a] However, by monitoring ternary complex formation with radiolabeled eIF-2 as well as radiolabeled Met-tRNA$_f$, it was observed that eIF-2 exhibits 100% activity in ternary complex formation, when its concentration is kept above a critical level, and the ratios of GTP to GDP and of Met-tRNA$_f$ to eIF-2 are rigorously controlled (*133*). In addition, eIF-2 is subject to losses, particularly at low concentrations, caused by adsorption to the wall of the reaction vessel (*133*). These losses can be prevented by the addition of proteins with characteristics similar to co-eIF-2A (*133*). Thus, although

---

[2a] E.g., co-eIF-2A (*124–126*) and ESP (eIF-2 stimulating protein) (*127–132*).

the possibility remains that the proportion of eIF-2 entering ternary complexes is influenced by "regulatory" proteins, a rigorous demonstration of their existence has not been made, and such a demonstration will require evaluation of the potential contribution from the artifacts outlined above.

In addition to these difficulties, monitoring of the steady-state level of ternary complex formation provides only limited information; assessment of conditions that influence the rate of the reaction is likely to be more illuminating. In addition, the assessment of eIF-2 activity solely by its ability to take part in ternary complex formation overlooks the importance of its capacity to function catalytically.

3. Ternary Complex Binding to 43 S_N Ribosomal Subunits

Met-tRNA$_f$ binding to the 43 S_N ribosomal subunit occurs exclusively via the ternary complex (eIF-2) · GTP · (Met-tRNA$_f$) (75, 96, 134). The resulting 43 S preinitiation complex has the form (eIF-2) · GTP · (Met-tRNA$_f$) · 43 S_N (77, 96, 97, 134). Studies with purified components demonstrate that although eIF-2 is sufficient for the binding of Met-tRNA$_f$ to 43 S_N ribosomal subunits, the binding takes place more readily in the presence of eIF-3 (96). At saturating levels of eIF-2, eIF-3 stimulates Met-tRNA$_f$ binding from 50 to 100% (96). In the presence of limiting amounts of eIF-2, eIF-3 stimulates two- to fourfold (96).

As a further demonstration of the interdependence between eIF-2 and eIF-3, the interaction of eIF-3 with the 40 S · ternary complex is more stable than its binding to 40 S ribosomal subunits alone (96). The necessity for bound eIF-3 is supported by *in vivo* studies, since only 40 S subunits associated with eIF-3 are found bound to Met-tRNA$_f$ (135–137). All subunits of eIF-2 and eIF-3 are found bound to the 43 S preinitiation complex (74, 96, 97, 99). eIF-1 and eIF-4C have also been reported to have stimulatory effects on Met-tRNA$_f$ binding to 43 S_N ribosomal subunits (134, 138). eIF-4C seems to stabilize Met-tRNA$_f$ binding, as the factor reduces the rate of exchange of Met-tRNA$_f$ in 43 S preinitiation complexes (138). The binding of Met-tRNA$_f$ to 43 S_N ribosomal subunits in eukaryotes is independent of mRNA (70, 96, 97, 136). In reticulocyte lysate, Met-tRNA$_f$ is normally only found in association with 43 S_N ribosomal subunits in which no mRNA can be detected (89, 134, 137, 139).

4. Additional Cofactors

There are reports of other factors and cofactors, in addition to co-eIF-2A and ESP, that affect ternary complex formation and Met-tRNA$_f$ binding to 43 S_N ribosomal subunits. Co-eIF-2B, a cofactor also referred to as EIF-2 and TDF, is a factor that promotes dissociation of ternary complexes formed at high Mg^{2+} concentrations, but that stimulates Met-tRNA$_f$ binding to 43 S_N ribosomal subunits (140–142). Co-eIF-2C, also referred to as EIF-2B, is a factor that reverses the observed Mg^{2+} inhibition of ternary complex formation (140).

It is not yet possible to appreciate the physiological role or significance of these cofactors because of real or apparent differences in the characteristics of eIF-2 prepared in those laboratories that report cofactor requirements. These differences in eIF-2 are observed as differences in activity in ternary complex formation and in Met-tRNA$_f$ binding to 43 S_N ribosomal subunits. The eIF-2 isolated by some is very active in ternary complex formation (10 to 100%) in the absence of cofactors (64, 97, 99, 133, 134). The rate of ternary complex formation is insensitive to elevated Mg^{2+} and is very rapid, the maximum extent of binding being achieved in less than a minute, even at 2–4°C (64, 97, 107). In addition, eIF-2 with these characteristics is sufficient for Met-tRNA$_f$ binding, although eIF-3 will provide some additional stimulation (96, 99).

In contrast, eIF-2 prepared by others exhibits low activity in ternary complex formation (< 10%) (124, 127, 131, 142, 143) and is sensitive to Mg^{2+} (131). It requires a cofactor (ESP, co-eIF-2A) to confer higher activity (20–70%) (124, 127). In the presence of this cofactor, the rate of complex formation, although slow (even at 37°), is linear for approximately 15 minutes (124, 142). Thus, eIF-2 has a requirement for additional factors (e.g., co-eIF-2B) for Met-tRNA$_f$ binding to 43 S_N ribosomal subunits (126).

Assessment of the role of these cofactors first requires evaluation of the differences in the reported characteristics of eIF-2. It is possible that some of the differences arise from the Millipore filtration binding assay of ternary complex formation. Past experience with other systems suggests that this type of assay may generate artifacts.

D. Binding of mRNA to 43 S Preinitiation Complex

1. MESSENGER RNA BINDING PROTEINS

Binding of mRNA to the 43 S preinitiation complex, (eIF-2) · GTP · (Met-tRNA$_f$) · 43 S_N, is the third major step in the initiation of protein synthesis, depicted in Fig. 4. Work on reconstituted, fractionated, and unfractionated systems has demonstrated that mRNA binding to 40 S ribosomal subunits is totally dependent on bound Met-tRNA$_f$ and eIF-2 (97, 134). In fractionated, reconstituted systems, three initiation factors (eIF-3, eIF-4A, eIF-4B) are required

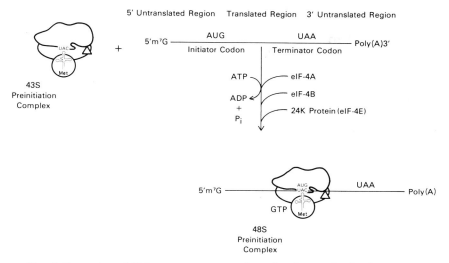

FIG. 4. Formation of 48 S preinitiation complex. This depicts the third step in initiation, the interaction of a 43 S preinitiation complex with mRNA, an ATP-requiring process. The involvement of initiation factors during this step is not fully resolved, because of the contaminating 24,000-dalton mRNA binding protein in eIF-3 and -4B. Messenger RNA bound to the preinitiation complex is associated with characteristic proteins, the functions of which are unknown.

for this step, together with ATP (64, 78, 97, 134). Omission of any of the initiation factors results in a 10- to 20-fold decrease in either the rate of globin synthesis or in the extent of mRNA binding to the 43 S preinitiation complexes (97, 134). Factor eIF-3, which binds first to 40 S subunits together with eIF-4C, is discussed in a preceding section. Factor eIF-4A is a single polypeptide chain of molecular weight 48,000–50,000, previously identified as IF-EMC (145), IF-M4 (64), and IF-E4 (71, 78). It is an acidic protein with an isoelectric point of 5.8. Factor eIF-4B consists of a single polypeptide chain of molecular weight 80,000 that corresponds to IF-M3 (67) and IF-E6 (71, 78). The role of these factors appears to be related to a characteristic structural feature of all eukaryotic mRNAs. Most eukaryotic cell and viral mRNAs have a 5'-terminal structure of the form $m^7G(5')ppp(5')Nm-N'$ . . . , in which the terminal 7-methylguanosine is linked by a 5'-5' triphosphoric bridge to an adjacent methylated nucleotide. Several roles for such "caps" have been proposed, but the greatest attention has been paid to their influence on the translation of mRNAs (reviewed in *146–148*)).³ The efficiency of ribosome binding and translation of mRNA in cell-free protein-synthesizing systems appears to depend on the presence of cap structures, particularly in extracts from wheat germ or brine shrimp. However, the presence of a cap is not obligatory (*147, 148*); it appears to increase the affinity of mRNA for the small ribosomal subunit, and it is recognized during initiation (*149*).

The best demonstration of the involvement of eIF-4B in cap binding *in vivo* has come from studies on protein synthesis following poliovirus infection of tissue culture cells, which redirects protein synthesis from host–cell polypeptide formation to virus-specific products (*150, 151*). The translation of capped mRNAs decreases in extracts of poliovirus-infected HeLa cells, whereas poliovirus mRNA, which has no cap, continues to function well *in vitro* (*82, 152*). Crude initiation factor preparations from poliovirus-infected cells support the initiation of viral, but not host cell, mRNA translation with host cell ribosomes (*151, 153*). In addition, the addition of eIF-4B to infected cell extracts restores the translation of capped mRNA without increasing poliovirus synthesis (*152*).

Eukaryotic initiation factors have been assayed in fractionated systems for cap-binding activity by the retention of mRNA–protein complexes on nitrocellulose filters, showing that eIF-2, eIF-3, and eIF-4B bind to cap structures (*81, 154*). However, the binding of all factors except eIF-3 is decreased by 18 S rRNA and by uncapped or unmethylated reovirus RNA, suggesting that the binding is nonspecific (*1*).

Cap-specific interactions cannot be assumed simply because m⁷G nucleotides inhibit binding of initiation factors, since, by the filter method, m⁷G nucleotides (or a contaminant of m⁷G nucleotide preparations) inhibit nucleoprotein complex formation nonspecifically (*155*). Sonenberg *et al.* addressed this problem by treating purified initiation factors with [³H]methyl-labeled reovirus mRNA oxidized to convert the 5′-terminal m⁷G to a reactive dialdehyde (*81*). Complexes between mRNA and initiation factors were reduced with $NaBH_3CN$ to stabilize the complex formed between the oxidized cap and neighboring free amino acid groups of bound proteins. The polypeptides that were ³H-labeled as a consequence of cross-linking to the cap were identified by electrophoresis with dodecyl sulfate on polyacrylamide gels and fluorography. From these studies it appears that the "cap-binding activity" is associated with a 24,000 dalton polypeptide pres-

---

³ See also Vol. 19 (1976) of this series, containing many articles dealing with "caps" and other features of mRNA. [Ed.]

ent as a minor contaminant in eIF-3 and eIF-4B (81). This finding casts doubt on the involvement of eIF-3 and eIF-4B in cap binding. It must be appreciated that eIF-4B has only been defined as an activity in reconstituted fractionated systems, and it is not yet clearly established whether the predominant peptide in eIF-4B has mRNA binding properties of its own, or whether that activity resides solely in the contaminating 24,000-dalton peptide.

Because this polypeptide is present in less than stoichiometric amounts, it should be regarded as a distinct initiation factor, and perhaps classified as eIF-4E. It has been purified by affinity chromatography, using the levulinic acid $O^{2,3}$-acetal of m^7G bound to Sepharose (83). The affinity-purified 24,000-dalton cap-binding protein partially restores the translation of capped mRNAs *in vitro* without increasing noncapped mRNA translation (84). This further strengthens the possibility that the "restoring activity" and, presumably, cap-binding properties previously observed with eIF-4B are due to this 24,000-dalton polypeptide.

This peptide copurifies with a fraction that stimulates the activity of *decapped* viral mRNA (82) and has led to the suggestion that the 24,000-dalton cap-binding protein recognizes not only the 5'-terminal cap but also sequences adjacent to it.

Controversy has arisen over whether or not eIF-2 can bind mRNA. A few investigators have reported that the factor can associate with poly(A) and globin mRNA (156–159). Also, it is suggested that the β subunit of eIF-2 is a component of reticulocyte mRNPs (159, 160), although careful studies have failed to demonstrate any specific binding of eIF-2 to cap structures of mRNA, even using radiolabeled $m^7GppGm-C$ (81). Factor eIF-2 does bind tightly at internal sites of mRNAs, and it is possible that this binding serves some function; a factor that binds both GTP and Met-tRNA$_f$ may be expected to have an affinity for many RNA molecules, and eIF-2 is also found to bind to 18 S rRNA, uncapped reovirus RNA, and double-stranded RNA (81). However, eIF-2β is not identical with the similar peptide found in reticulocyte mRNPs (161); they differ in their molecular weights, isoelectric points, and the products they yield upon protease or cyanogen bromide cleavage (161).

2. USE OF AUG

In reconstituted, fractionated systems, mRNA can be replaced in some investigations by the initiator codon, AUG (162). Although an 80 S initiation complex can be formed in the presence of AUG, the characteristics of the complex and the conditions necessary for its formation differ. For instance, the binding of AUG to a 43 S preinitiation complex occurs in the absence of ATP or of eIF-4A and -4B, which are required for natural mRNA binding (163). In addition, in the absence of 60 S ribosomal subunits, eIF-5 stimulates GTP hydrolysis and initiation factor release from 48 S preinitiation complexes formed with AUG (99, 164). By contrast, 48 S preinitiation complexes formed with mRNA require the presence of 60 S ribosomal subunits in addition to eIF-5 for GTP hydrolysis and factor release (164). Finally, in 80 S initiation complexes formed with AUG instead of mRNA, Met-tRNA$_f$ becomes accessible to the Met-tRNA$_f$ deacylase bound to the 43 S_N subunit (105).

3. Energy Requirements for mRNA Binding

ATP is required during mRNA binding to the 43 S preinitiation complex. This was first observed in a wheat germ cell-free system (165) and later confirmed in a fractionated mammalian system (70, 71, 134). During mRNA binding, ATP is hydrolyzed; nonhydrolyzable ATP analogs do not sustain mRNA binding. In a reconstituted, fractionated system, ATP is hydrolyzed nonstoichiometrically, although this may not be the case *in vivo* (134). Whether ATP is hydrolyzed per se or utilized in a phosphorylation reaction is not clear. However, fluoride, an inhibitor of ATPases, inhibits initiation after mRNA binding, which suggests that the ATP may be required for a phosphorylation reaction (89). If ATP is required for a phosphorylation reaction, it is possible that either eIF-4A or -4B is the phosphoric group acceptor. In support of this contention, it should be noted that when 48 S preinitiation complex formation occurs in the presence of AUG, rather than mRNA, eIF-4A, -4B, and ATP are no longer required (163). In addition, eIF-4B is phosphorylated in cultured reticulocytes (166) and in reticulocyte lysate (74).

4. Recognition of mRNA during Initiation

The nucleotide sequences of many eukaryotic mRNA initiation regions are now known, and they possess few common features (144, 149). Each has an AUG codon preceded by a nontranslated region, and a 5' terminal cap structure (149). However, the region between the cap and the AUG is highly variable, both in length and composition (149, 167). Also, the evidence for rRNA · mRNA base-pairing in eukaryotes is weak (168). The cap structure by itself is not sufficient to form a stable initiation complex (169). The base composition of the RNA fragment adjacent to the cap structure affects the relative affinity of the 5' fragment for ribosomes (170, 171). Apart from the m^{7}G substituent, the 2'-O-methyl group of the penultimate nucleotide enhances the extent of ribosome binding to cap-containing synthetic polymers (170). The m⁷G(5')ppp(5')Am caps seem to bind more efficiently than m⁷G(5')ppp(5')Gm caps (172).

Several attempts have been made to obtain information about the sequences of ribosome binding sites in mRNAs by subjecting initiation complexes to nuclease digestion and isolating those fragments protected by the ribosomes. Such studies have shown, with both globin (173) and reovirus mRNA (169), that the sequences protected by 40 S complexes are longer than those protected by 80 S complexes, and extend further towards the 5' end. Each 80 S-protected sequence has an AUG initiation codon, and is a subset of the 40 S-protected sequence from the same message (169). The importance of the 5'-terminal sequences was shown by the greater capacity of 40 S-protected sequences to rebind to the small ribosomal subunit. The limited sequence homology led to the proposal of a minimal recognition mechanism for the initiation of eukaryotic mRNAs, in which 43 S preinitiation complexes bind at or near the 5' cap structure and then move along the RNA until an AUG is encountered, which will be complementary to the anticodon on the tRNA (149, 169, 174). This hypothesis is substantiated by the inability of eukaryotic ribosomes to bind to circular mRNA, which demonstrates the need for a free 5' end (175). This is in contrast to prokaryotic ribosomes, which are able to translate circular messages, because initiation involves base-pairing between complementary regions at the 5' region of the mRNA and the 3' end of 16 S ribosomal RNA (176).

Two predictions from this model are that internal AUG triplets will not be used as initiation sites, and that eukaryotic ribosomes will be unable to translate cistrons other than the 5'-proximal one in polycistronic mRNAs. These predictions are fulfilled by the observation that certain prokaryotic mRNAs, when modified at their 5' terminal with a cap structure, are translated in a eukaryotic cell-free system as efficiently as eukaryotic mRNA (177). However, only the first gene in an operon is translated (178).

5. Evidence Supporting the Existence of mRNA-Containing Preinitiation Complexes *in Vivo*

Although a preinitiation complex containing both Met-tRNA$_f$ and mRNA can readily be formed *in vitro* (97, 134), it is difficult to detect *in vivo* in the presence of 60 S ribosomal subunits. This suggests that mRNA binding to 43 S preinitiation complexes constitutes a rate-limiting step in initiation under normal conditions (87). However, preinitiation complexes containing stoichiometric amounts of mRNA accumulate in reticulocyte lysate in the presence of the inhibitors of initiation, edeine and potassium fluoride, both of which inhibit 60 S subunit joining (87–89).

The preinitiation complexes that accumulate have a higher sedimentation coefficient (48 S) in sucrose gradients than the 43 S preinitiation complexes (87–89). In the presence of edeine, the 48 S preinitiation complex contains stoichiometric amounts of mRNA, Met-tRNA$_f$, eIF-2, and eIF-3 (87, 88), although eIF-4A and -4B cannot be detected unambiguously (see Fig. 5). In a wheat germ cell-free system, in the presence of edeine, the 40 S subunit appears to move past the AUG codon (179). This may result in the release of mRNA binding factors. However, the doubt cast on the role of eIF-4B may provide a sufficient explanation for its absence.

In the presence of fluoride, a 48 S complex accumulates that contains stoichiometric amounts of mRNA and the deacylated tRNA$_f$ (87–89, 180). This defective preinitiation complex does not contain eIF-2 or eIF-3 (181). It is speculated that, in the absence of 60 S subunit joining, the Met-tRNA$_f$ is hydrolyzed by the Met-tRNA$_f$ hydrolase known to be associated with 43 S$_N$ ribosomal subunits (101–104). It is also suggested that, after hydrolysis of Met-tRNA$_f$, the binding of eIF-2 and eIF-3 becomes unstable, and the factors are released. Since both types of mRNA-containing preinitiation complexes sediment at 48 S, it is clear that the increase in sedimentation coefficient does not reflect an increase in molecular weight caused by the presence of additional factors, but may simply reflect a change in conformation.

6. Messenger RNA Exists as Messenger Ribonucleoprotein Complexes

Messenger RNA molecules are found in cell cytoplasm as ribonucleoprotein complexes (mRNPs) (182–185). A comparison of mRNPs found free in cytoplasm or bound in polysomes demonstrates differences in buoyant density due to differences in protein content (186–189). For instance, in duck immature erythrocytes, globin mRNA is found free in the cytoplasm as 20 S mRNPs and as 15 S mRNPs in polyribosomes (186, 187). Examination of cytoplasmic-free and polysomal mRNP complexes from a variety of eukaryotic cells, rabbit reticulocytes, ascites tumor cells, rat liver, and chick embryo muscle cells, indicates that a set of similar proteins of molecular weights 44,000, 52,000, 58,000, 65,000, and 78,000 are present in a wide variety of cells (190).

Proteins bound to the poly(A) sequence, the 65,000 and 78,000 dalton proteins, are present in both cytoplasmic-free and polysomal

INITIATION OF MAMMALIAN PROTEIN SYNTHESIS 149

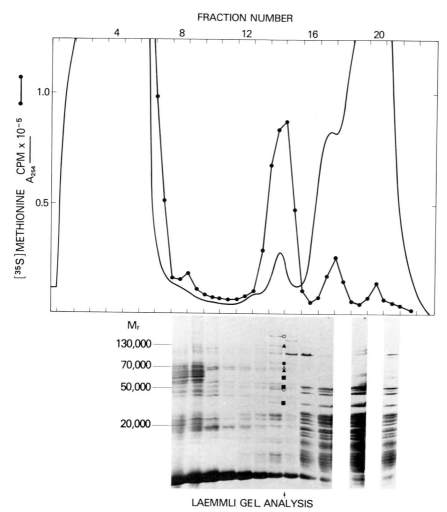

FIG. 5. Composition of 48 S preinitiation complex. The top half of the figure depicts the distribution of [^{14}C]eIF-2, [^{35}S]Met-tRNA$_f$, and mRNA in edeine-inhibited lysate in which 48 S preinitiation complexes accumulate. The amount of mRNA accumulating in the 48 S complex represents approximately 60% of the mRNA contained in lysate. This may be compared with the effects of hemin deprivation seen in Fig. 7, where only approximately 10 → 15% of mRNA accumulates in preinitiation complexes. The proteins contained in each gradient fraction are shown in the lower half of the figure, resolved on 15% Laemmli polyacrylamide gels. The Coomassie-blue-stained polypeptide patterns are aligned beneath the corresponding gradient fraction. The α, β, and γ subunits of eIF-2 are indicated by ■; other polypeptides associated with the 48 S preinitiation complex have been identified as subunits of eIF-3, ▲; other proteins that characteristically appear (●, ○, □, △) may represent mRNP proteins and/or other initiation factors.

mRNPs (*188, 189, 192–193*). The 78,000 dalton protein is also present in mRNPs newly released from the nuclei, which suggests that this protein becomes bound to the 3' poly(A) tracts of hnRNA, and subsequently remains in association in the cytoplasm (*192, 194, 195*). This is consistent with its identification as poly(A) polymerase (*196*).

In addition, a set of specific proteins are bound to the nonpoly(A) sequence of mRNA in free cytoplasmic mRNP (*190*). These are released during polysome formation (*190*). Many of the proteins usually associated with free cytoplasmic mRNPs are absent in mRNPs newly released from the nucleus. This suggests that there is a cytoplasmic pool of mRNP proteins, the size and availability of which may serve a regulatory function.

Although deproteinized mRNA is capable of taking part in protein synthesis in cell-free extracts, this occurs via interaction with the cytoplasmic pool of mRNP proteins. It is not known to what extent the composition of free cytoplasmic mRNP affects the subsequent utilization of mRNA, since reconstituted, fractionated systems have not been used to evaluate the role of mRNP proteins and their relocation during translation.

Unequivocal evidence is not available to demonstrate that some mRNP proteins are initiation factors (see also Section II,C).

7. MESSENGER RNA SPECIFICITY FACTORS

Suggestions that there are mRNA specificity factors altering the rates of translation of specific mRNAs have provoked considerable controversy (*1b–3, 197–203*). Current understanding favors the view that the existence of mRNA-specific factors is not proved, and that such explanations are not necessary to account for observed differences in the translation of different mRNAs (*2, 3*). Differences in affinities between various mRNAs for 43 S preinitiation complexes or initiation factors, or both, lead to changes in relative translational efficiencies (*1b–3*). For instance, a nonspecific reduction in the rate of polypeptide chain initiation at or before mRNA binding will result in the preferential inhibition of translation of mRNAs, with lower rate constants for polypeptide chain initiation (*1b–3*). Such changes do not require changes in the specificity of the protein synthetic apparatus for various mRNAs, and suggest that the postulation of mRNA-specific factors is superfluous; it is necessary only to postulate that the factors are not mRNA-specific, but rather affect the translation of all mRNAs (*2, 3*). Predictions from kinetic modeling studies also indicate that there is a far greater potential for the selective increase or reduction of the translation of mRNAs with low affinities for the relevant mRNA-binding factor(s) by changes in the activity of these factors than by reducing the rate of formation of 43 S preinitiation complexes (3).

The initiation factors reported to possess mRNA-discriminatory properties are eIF-3 and eIF-4B (200–203). Factor eIF-4B is required in different amounts for the translation of encephalomyocarditis and globin mRNAs (200). Also, β-globin mRNA has a 50-fold greater affinity than α-globin mRNA for eIF-4B (201). Thus, differences in the affinity of eIF-4B for different mRNAs may provide a mechanism for the differential control of protein synthesis. By contrast, proteins associated with muscle eIF-3 enable myosin mRNA to compete more efficiently with globin mRNA in a cell-free protein synthesis system (202, 203).

The observations on eIF-3 and -4B reported above may reflect the requirement for the 24,000-dalton protein involved in cap binding (described in Section II,D,1), which is found as a contaminant of eIF-3 and -4B (81, 82, 84).

E. 60 S Ribosomal Subunit Joining and Initiation Factor Release

The fourth step during 80 S initiation complex formation is the joining of the 60 S ribosomal subunit to the 48 S preinitiation complex depicted in Fig. 6. This step is mediated by eIF-5 (67, 134) and is accompanied by loss of eIF-2 and eIF-3 (96, 99, 134, 164, 204). The initiation factor eIF-5 is a large single polypeptide of molecular weight 125,000, variously reported in the literature as IF-M2A (67), IF-L2 (205), IF-E5 (71), IF-3 (110), IF-11 (206), F-0.25 (207), and IF-E2 (207a). This factor is reported to be phosphorylated in reticulocyte lysate (74). The 80 S initiation complex formation is another step stimulated by eIF-4C (100, 134, 138, 164). In a reconstituted, fractionated system utilizing AUG as template, eIF-4C increases the proportion of 48 S preinitiation complexes that form 80 S initiation complexes, although it is not sufficient to stimulate 80 S formation in the absence of eIF-5 (96).

Immediately before 60 S subunit joining, the 48 S preinitiation complex contains equimolar amounts of 40 S ribosomal subunits, eIF-2, eIF-3, eIF-4C, Met-tRNA$_f$, GTP, and mRNA; stable binding of eIF-4A and -4B has not been demonstrated. Although initiation can proceed as far as the formation of the 48 S preinitiation complex with nonhydrolyzable GTP analogs, both initiation factor release and 60 S ribosomal subunit joining require GTP hydrolysis (96, 99, 134, 164, 204). The GTP initially bound to eIF-2 prior to ternary complex formation is the same molecule that is hydrolyzed to permit 60 S subunit joining to occur (208). It is experimentally difficult to determine

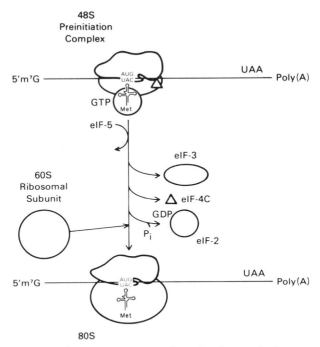

FIG. 6. Formation of 80 S initiation complex. This depicts the last step in initiation, the joining of the large ribosomal subunit to the 48 S preinitiation complex. During this step, which is catalyzed by eIF-5, GTP is hydrolyzed, and eIFs −2, −3, and −4C are released. Factor eIF-2 is released as inactive eIF-2 · GDP, which is not immediately available for reutilization.

whether initiation factor release precedes, or is the result of, 60 S subunit joining. However, since the two factors, eIF-2 and eIF-3, amount to more than 50% of the mass of the 40 S ribosomal subunit, a model incorporating factor release before subunit joining has been favored. In the presence of GTP and eIF-5, omission of 60 S ribosomal subunits does not prevent GTP hydrolysis or factor release when AUG is used to replace mRNA (164).

Although release of eIF-2 and eIF-3 has been demonstrated, it is unclear whether they are released in a form capable of reinitiating protein synthesis. Reinitiation does not occur in 80 S initiation complex formation in reconstituted, fractionated systems. It is speculated that eIF-2 is released as an inactive (eIF-2) · GDP complex following the hydrolysis of GTP in the 48 S complex. Since 100-fold excess in the concentration of GTP is required to replace bound GDP on eIF-2, it is probable that the eIF-2 released is not immediately available for reutilization. This implies a need for either a recycling factor that could displace bound GDP (analogous to EF-Ts) (208a), or facilitate GDP/GTP exchange.

III. Regulation of Initiation

A. Regulation of Globin mRNA Translation by Hemin

1. Historical Introduction

The most intensely investigated animal cell system exhibiting translational control is the mammalian reticulocyte. The reticulocyte is the last stage in red blood cell development before maturation to the erythrocyte. More than 90% of the protein produced by reticulocytes is globin. Kruh and Borsook first reported that heme and globin synthesis proceed at parallel rates (209). This led to speculation that there is a mechanism for coordinating the synthesis of globin with that of its prosthetic group, heme.

Globin synthesis in intact reticulocytes is closely coupled to the availability of ferrous ions or hemin (209a, 210). During deprivation of Fe^{2+}, protein synthesis in reticulocytes declines, and polysomes disappear after a lag period during which normal rates of protein synthesis and normal polysome distributions are maintained (210–212). These changes, characteristic of a state of inhibited initiation, can be reversed by the addition of Fe^{2+} or hemin (209a, 210).

There are other examples of the coordinated synthesis of heme and globin. In chick blastoderm, where δ-aminolevulinic acid synthesis is rate-limiting for heme production, addition of this acid to the culture results in the earlier production of hemoglobin by 6 to 12 hours (213). In Friend-virus-infected mouse erythroleukemia (MEL) cells, after induction with Me_2SO, globin mRNA is available after 2 days, although hemoglobin is not manufactured until iron chelatase is synthesized, beginning on the fourth day (214). In contrast, there are examples demonstrating a lack of coordinate control between heme and globin synthesis. For instance, Co^{2+} inhibits the synthesis of heme, but allows globin synthesis to proceed normally (215). Similarly, there is an MEL-cell line, FW, that has a deficiency in heme synthesis (216). Because heme is not produced in significant quantities, hemoglobin does not accumulate in these cells after treatment with inducers, although globin is synthesized after induction.

It is not yet known whether the coordinate control of heme and globin synthesis in reticulocytes is representative of a general regulatory mechanism for the synthesis of other hemoproteins, such as tryptophan pyrrolase, cytochrome P-450, or catalase, although hemoproteins are commonly formed in a molar ratio of one heme to one apoprotein, with no excess of either component accumulating in the cell (217). The presence of rapidly turning-over heme in liver and bone marrow indicates that heme is constantly being made in excess of the requirement for hemoprotein synthesis (218). This suggests that the maintenance of molar ratios is usually regulated at the level of turnover of the prosthetic group. However, the accumulated data on hemoglobin synthesis suggest that the control of apoprotein production may be important in circumstances where heme synthesis is interrupted.

2. RETICULOCYTE LYSATE

a. Kinetics of Inhibition. Investigation of the mechanisms underlying the control of protein synthesis by heme was accelerated by the development of a cell-free protein-synthesizing system that retains characteristics similar to those of intact reticulocytes. The preparation does not contain mitochondria and, therefore, does not synthesize heme. Reticulocyte lysate, appropriately supplemented, synthesizes globin at rates approaching those observed in intact reticulocytes (219, 220). The protein synthetic characteristics of reticulocyte lysate are depicted in Fig. 7. In the presence of hemin, a linear rate of translation, approaching 1 mol of globin per mole of 80 S[4] per minute, continues for up to 120 minutes (88). Optimal concentrations of hemin vary from 20 to 40 μM, but higher concentrations are inhibitory (219). The iron-free compound, protoporphyrin IX, is inactive in the maintenance of protein synthesis (219).

In the absence of hemin, protein synthesis proceeds at the same rate as in its presence for 4–6 minutes and then declines abruptly to less than 10% of the initial rate (219, 220). The onset of inhibition is accompanied by disaggregation of polysomes, diagnostic of a decreased rate of initiation (219, 220). The length of the lag period is inversely proportional to the incubation temperature, or to the rate of translation before shutoff (221). This observation led to the idea that a component of the system required for initiation is progressively inactivated or exhausted. The inhibition can be reversed if hemin is added soon after the start of incubation, although its effectiveness in reversing the inhibition is diminished with increasing time of incuba-

---

[4] 80 S ribosome equivalents.

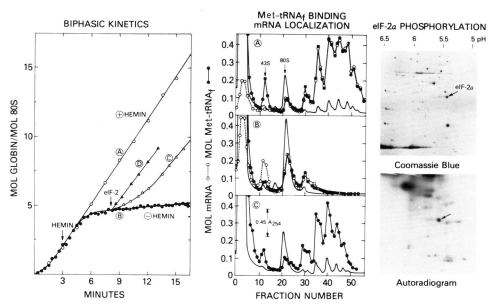

FIG. 7. Hemin-regulated translational inhibition. The characteristics of protein synthesis in rabbit reticulocyte lysate are depicted. In the presence of hemin, globin synthesis occurs at a linear rate, approaching 1 mol of globin per mole of 80 S^4 per minute for up to 2 hours. Sucrose density gradient analysis of incubated lysate sampled at Ⓐ shows that most of the ribosomes are present as polysomes active in protein synthesis. In the absence of hemin, protein synthesis proceeds at the same rate as in its presence for 4 min, then declines abruptly to less than 10% of the initial rate. At this time, Ⓑ, the polysomes are disaggregated and the level of Met-tRNA$_f$ bound to 43 S_N ribosomal subunits is reduced to approximately 30%. However, the level of mRNA bound in this region is disproportionately high, indicative of the degeneration of preinitiation complexes. At this time, eIF-2α kinase is active and 20% to 30% eIF-2α is phosphorylated. If hemin is added back at 3 minutes, protein synthesis recovers, after a lag, to its initial rate, Ⓒ. This is reflected in the polysome profiles. If eIF-2 is added, protein synthesis recovers immediately, but transiently, to its initial rate, Ⓓ.

tion in its absence (222–224). The disaggregation of polysomes, which reflects a decrease in the rate of initiation relative to elongation, is accompanied by a marked diminution in the level of 43 S preinitiation complexes (225). In the absence of hemin, the level of Met-tRNA$_f$ bound to 43 S_N ribosomal subunits declines from the second to the fifth minute of incubation (225). This observation focused attention both on the formation of 43 S preinitiation complexes as the potential rate-limiting step in initiation of hemin-deficient lysates, and on eIF-2, the factor responsible for the binding of Met-tRNA$_f$ to 43 S_N ribosomal subunits. The inhibition of protein synthesis in the absence of hemin can be reversed by the addition of eIF-2 (99, 226, 227). Such observations gave rise to the speculation that eIF-2 is the lysate component progressively inactivated or exhausted during the lag period.

b. *Sites of Lesions in Initiation.* The initial observations on 43 S preinitiation complex formation, in the presence or the absence of hemin, came from measurements of the steady-state level of this complex, rather than the rate of its formation. The decreased binding of Met-tRNA$_f$ to 43 S$_N$ ribosomal subunits in the absence of hemin might have represented a decrease in the rate of 43 S preinitiation complex formation. However, it might also have represented an increase in complex dissociation, or the deacylation of Met-tRNA$_f$ in the 43 S preinitiation complex, or both.

Attempts have been made to differentiate between these possibilities in various ways. For instance, inhibitors of elongation (e.g., cycloheximide) cause the accumulation of 43 S preinitiation complexes, once the endogenous mRNA pool has been exhausted. Balkow *et al.* (228) evaluated the competence of 43 S preinitiation complexes, accumulated in the presence or the absence of hemin, to take part in 80 S initiation complex formation. The addition of AUG promoted a quantitative shift of bound Met-tRNA$_f$ to 80 S initiation complexes in cycloheximide-treated, hemin-supplemented lysate. However, in cycloheximide-treated hemin-deficient lysate, the transfer did not occur. This suggested that 43 S preinitiation complexes can accumulate in hemin-deficient lysate, but are unable to complete the initiation sequence. In addition, although hemin-deficiency resulted in a decreased binding of Met-tRNA$_f$ to 43 S preinitiation complexes, the levels of tRNA$_f$ bound remained essentially constant. This led to the proposal (228) that the decrease in protein synthesis in hemin-deficient lysate involves deacylation of Met-tRNA$_f$ already bound to 43 S preinitiation complexes, which could either be caused by an increased Met-tRNA$_f$ hydrolase activity, for which there is no evidence, or occur secondarily from a decreased utilization of preinitiation complexes.

Although a new lesion in preinitiation complex formation was thus identified, the data (228) do not exclude the possibility that 43 S preinitiation complex formation is also directly affected by hemin-deficiency. In this context, it should be remembered that when mRNA is replaced by AUG in deacylase assays, Met-tRNA$_f$ in the resulting 80 S initiation complexes is more susceptible to deacylase activity (see Section II,D,2). This suggests that the extent of the lesion in 43 S preinitiation complex utilization may be overestimated in assays using AUG rather than natural mRNA. The 43 S preinitiation complexes that accumulate in the presence or in the absence of hemin are capable of forming 80 S initiation complexes with globin mRNA (229).

Unfortunately, comparisons of the rates of 43 S preinitiation complex utilization, or initiation factor turnover, are not simple because of pool size differences; inhibitors of elongation trap more mRNA and ribosomes in polysomal structures in hemin-supplemented lysate, which results in the accumulation of fewer available ribosomal subunits.

Nevertheless, other work (230) supports the earlier observation (228) that preinitiation complexes can accumulate in hemin-deficient lysate. However, no definitive quantitative studies have yet been carried out to compare the rate of 43 S preinitiation complex formation in hemin-supplemented and hemin-deficient lysate. This is theoretically possible in unfractionated lysate utilizing inhibitors to block initiation at particular steps, or elongation. However, technical difficulties arise because the pool size of ribosomes that do not participate in protein synthesis is lower in hemin-supplemented than in hemin-deficient lysate. Also, the inhibitors of elongation of 60 S subunit joining that could be utilized may themselves have some direct effect on 43 S preinitiation complex formation.

The problem has been tackled using ribosomes isolated from lysate by filtration using Sepharose-6B (231). This study indicates that 43 S preinitiation complex formation is directly reduced in the absence of hemin. However, because the rate of initiation of "6B-ribosomes" is significantly lower than that observed in unfractionated lysate, this work does not provide the quantitatively useful data necessary to evaluate the relative contributions that reduction in 43 S preinitiation complex formation and reduced complex utilization may provide. Probably one of the most pertinent observations to date is that although the rate of protein synthesis in hemin-deficient lysate is less than 10% of that observed in hemin-supplemented lysate, levels of 43 S preinitiation complexes are reduced only to 35–50%.

c. Messenger RNA Distribution during Hemin Deprivation. When 60 S subunit joining is blocked by edeine, a 48 S preinitiation complex accumulates that contains equimolar quantities of 40 S ribosomal subunits, Met-tRNA$_f$, mRNA, eIF-2, and eIF-3 (87–89). By contrast, fluoride, which also inhibits 60 S subunit joining, results in the accumulation of a 48 S preinitiation complex (or degradation product) containing equivalent amounts of 43 S$_N$ ribosomal subunits, mRNA, and tRNA$_f$ (89, 180). In addition to the deacylation of Met-tRNA$_f$, eIF-2 and eIF-3 are also lost (232). If 60 S ribosomal subunit joining is defective in hemin-deficient lysate, similar defective preinitiation complexes might be expected to accumulate. The contention that a 48 S preinitiation complex, containing mRNA, also accumulates during hemin deprivation (233) is not supported by the findings (Jagus et al., unpublished) of preinitiation complexes (or degradation products) with only a slightly larger sedimentation coefficient than 43 S. These complexes do not accumulate to the same levels as the 48 S complexes seen after edeine or fluoride (compare mRNA localization in Figs. 5 and 7). As illustrated in Fig. 7, the complexes contain mRNA, but disproportionately low levels of Met-tRNA$_f$ and initiation factors. These particles are considered as the degradation product of a defective preinitiation complex incapable of 60 S subunit joining. Bound Met-tRNA$_f$ is unmasked and susceptible to deacylase activity during 60 S subunit joining, an event inhibited by GTP (105). This observation leads to the speculation that deacylation normally occurs only if a defective initiation cycle takes place. Failure to form a functional complex on interaction with a 60 S ribosomal subunit could lead to hydrolysis of bound GTP, leaving the Met-tRNA$_f$ exposed to deacylase activity. The Met-tRNA$_f$ deacylase activity, could, therefore, represent a mechanism that allows the release and subsequent reutilization of initiation components trapped in defective preinitiation complexes. It is speculated that the same modification/inactivation of eIF-2 that results in a decrease in the rate of Met-tRNA$_f$ binding to 43 S_N ribosomal subunits during hemin deprivation also affects 60 S subunit joining, directly or indirectly.

3. INHIBITOR FORMATION

a. Initial Description. Concurrent with the accumulation of evidence that the inhibition of protein synthesis following hemin deprivation is caused by modification/inactivation of eIF-2, data were accumulating to show that the inhibition is a consequence of the activation and enzymic action of an inhibitory protein (223, 224, 234). An advantage of the use of reticulocyte lysate is that experiments can be carried out where uninhibited reticulocyte lysate is mixed with inhibited lysates; as a consequence, protein synthesis ceases, suggesting the presence of a dominant inhibitor in inhibited lysate (235, 236).

The inhibitor appears in the postribosomal supernatant of lysate incubated without hemin at temperatures above 25°C (223, 224, 237, 238) and requires neither ribosomes nor protein synthesis for its activation (234). In the presence of hemin, formation of the inhibitor is suppressed; consequently, it was termed the heme-controlled repressor (HCR) and also later the heme-regulated inhibitor (HRI). The inhibitor is formed in three stages (224, 234, 237). If hemin is added after short periods of incubation of the postribosomal supernatant in the absence of hemin (up to 45 minutes), the activated inhibitor reverts to an inactive form (224). After longer incubation times (45–150 minutes), addition of hemin alone to the postribosomal supernatant is no longer sufficient to cause a reversal of the inhibitor to an inactive form; however, incubation in the presence of hemin and lysate does cause reversal. Such evidence led Gross to propose the existence of an inactive proinhibitor that is converted to a hemin-reversible inhibitor in the absence of hemin (224). After longer incubation times (45–150 minutes), in the absence of hemin, this is converted to an intermediate form, inactivated in the presence of hemin by a lysate component but not directly by hemin. After very long incubation times (more than 150 minutes), there is an additional transition to an irreversible form (224). The inactive proinhibitor and HCR have similar molecular weights in crude preparations, suggesting that the conversion to HCR involves a conformational change (237). This proposal is supported by the finding that there is little immunological cross-reactivity between HCR and its precursor (239). A heat-labile factor is involved in the conversion of proinhibitor to HCR suggesting an enzymic step (234). Irreversible inhibitor is also formed very rapidly if postribosomal lysate supernatant is incubated with N-ethylmaleimide or o-iodosobenzoic acid, suggesting an involvement of sulfhydryl groups in the conversion (237). This suggestion is strengthened by the observation that dithiothreitol prevents the conversion of proinhibitor to an active form and inactivates the intermediate form of HCR, but not the final irreversible form (240).

The irreversibility of the final form enables it to be assayed by its effects on protein synthesis in hemin-supplemented lysate. Consequently, this form of HCR has received considerable attention, although it is not the physiologically active species.

b. *Characterization and Purification of the Inhibitor.* In crude lysates, the inhibition of protein synthesis caused by hemin deprivation can be prevented or reversed by high concentrations of cAMP and 2-aminopurine (241, 242). Since these are purine analogs, the effects of ATP and GTP on inhibition of protein synthesis and on inhibitor formation were investigated. ATP enhances inhibition and blocks the reversal by cAMP (241). Conversely, GTP prevents the decline in protein synthesis and reverses the inhibition caused by hemin deprivation (241). In a crude ribosome preparation of gel-filtered lysate used to monitor Met-tRNA$_f$ binding to 43 S_N ribosomal subunits, the hemin-controlled inhibitor is effective only in the presence of ATP (243). These observations evoked the speculation that the inhibitor functions as a protein kinase (243). The speculation was supported by the fact that nonhydrolyzable ATP analogs could not be substituted for ATP (243), and led to a search for ATP-dependent protein kinase activity in inhibitor preparations (244, 245). Two peptides become phosphorylated in crude ribosomal preparations, one of which is the α subunit of eIF-2; the other, of molecular weight 96,000, corresponds to one of the peptides found in inhibitor preparations (245). Additional reports confirmed that the hemin-controlled inhibitor contains a protein kinase activity that phosphorylates eIF-2α (246–251).

All preparations that inhibit protein synthesis in reticulocyte lysate exhibit protein kinase activity *in vitro*. This kinase activity is not stimulated by cAMP; in fact, cAMP and 2-aminopurine, which prevent the inhibition of protein synthesis, prevent phosphorylation of eIF-2α (245). By contrast, GTP appears to prevent activation of the kinase (243). Additionally, during incubation of lysate in the absence of hemin, endogenous eIF-2α becomes phosphorylated (249–251), although the extent of phosphorylation has not yet been directly quantitated. Such circumstantial evidence led to the hypothesis that the hemin-controlled inhibitor is a cAMP-independent protein kinase. It was speculated that activation of the kinase by incubation in the absence of hemin blocks the functioning of eIF-2 by phosphorylating it. The extent to which phosphorylation can account for the observed inhibition of protein synthesis is discussed in Section III,A,5.

The inhibitory kinase has been purified in its irreversible form (245, 247, 252, 253) and, recently, in an activated, but reversible, form (254). On purification of the "reversible inhibitor," the inhibitory activity is found in a single band after electrophoresis under nondenaturing conditions, which migrates as an 80,000 to 95,000 dalton peptide in dodecyl sulfate/polyacrylamide gels (254). At high concentrations, this polypeptide produces the characteristic biphasic kinetics of inhibition in reticulocyte lysate, and phosphorylation of eIF-2α. At lower concentrations, addition of the reversible kinase to lysate results in its inactivation; reversibility is dependent on the relative concentrations of free hemin and inhibitor (254).

Incubation of the purified "reversible kinase" with 20 μM hemin results in its inactivation, as judged by its inability to inhibit protein synthesis in lysate and a reduced capacity to phosphorylate eIF-2α (254). Since hemin is known to bind to many proteins, it is difficult to assess the physiological relevance of this finding. Also, it is difficult to identify this purified "reversible inhibitor" with any of the forms described by Gross (224), which makes it seem unlikely that this inhibitory kinase represents the physiological inactive or reversible form of the inhibitor.

Although there has been no unequivocal demonstration that phosphorylation of inhibitory kinase is directly responsible for its activation, the extent of phosphorylation observed nevertheless seems to be proportional to the amount of proinhibitor converted to active kinase (255). Furthermore, the fact that the inhibitor becomes phosphorylated with only the addition of ATP, at a rate insensitive to dilution of the inhibitor, is suggestive of an autophosphorylation reaction (254–255a). Gross calculates the incorporation of five moles of phosphate per mole of inhibitory kinase when the inhibitor is fully active (255).

An alternative mechanism for the catalytic activation of the inhibitory kinase, based on the possibility that it may be a substrate for a cAMP-dependent protein kinase, has been proposed (256–259). This model stems from the observation that a commercial preparation of bovine cardiac cAMP-dependent protein kinase, or its catalytic subunit, inhibits protein synthesis in reticulocyte lysate (256), and it was elaborated in the absence of any direct demonstration that the proinhibitor is a substrate for phosphorylation by exogenous cAMP-dependent protein kinase, or that inhibition of protein synthesis by cAMP-dependent kinase results in the phosphorylation of eIF-2α. Subsequent investigations by others with highly purified catalytic subunits of protein kinase II from reticulocytes (260, 261) and bovine heart kinase (261) failed to detect inhibitory effects on protein synthesis. Furthermore, although the addition of either of these catalytic subunits results in the phosphorylation of several polypeptides in reticulocyte lysate, neither the inhibitory kinase nor eIF-2α become phosphorylated (261). In addition, catalytic subunit inhibitor blocks the phosphorylation caused by the addition of catalytic subunits to lysate, but does not block the phosphorylation of the inhibitory kinase or eIF-2α that occurs during hemin deficiency (261). In summary, the available evidence excludes the involvement of a cAMP-dependent protein kinase in the regulation of protein synthesis by hemin.

4. PHOSPHORYLATION OF eIF-2α: ROLE OF eIF-2 PHOSPHATASE ACTIVITY

In the absence of hemin, endogenous eIF-2α rapidly becomes phosphorylated before the onset of translational inhibition (249–251). If hemin is then added, normal rates of protein synthesis are restored after a lag. Under conditions in which a constant specific activity of [γ^{32}P]ATP is maintained, no decrease in eIF-2α phosphorylation is discernible after the restoration of protein synthesis (250). However, under conditions optimized to monitor rapid phosphate turnover, a perceptible decrease in eIF-2α phosphorylation is discernible before protein synthesis is fully restored (251).

Although temporal relationships between the kinetics of translational inhibition and eIF-2α phosphorylation have been determined, the extent of phosphorylation during inhibition of protein synthesis has not been quantified by the measurement of [^{32}P]phosphate incorporation into eIF-2α.

It is necessary that covalent modification of a protein be readily reversible if it is to have regulatory significance; thus, it is not surprising that the steady-state level of eIF-2α phosphorylation is determined by a combination of highly active kinase and phosphatase activities (262). Although kinase activity appears to be regulated by hemin, hemin is without effect on the activity of the phosphatase. Furthermore, the addition of purified phosphatase to hemin-deficient lysate does not restore protein synthesis. This is because phosphatase activity in both hemin-supplemented and hemin-deficient lysate is extremely high. The addition of [^{32}P]eIF-2, labeled in both the α and β subunits, to reticulocyte lysate results in rapid dephosphorylation of eIF-2α ($t_{1/2} \simeq 20$ seconds) and occurs at the same rate in the presence or the absence of hemin (262). Since eIF-2 is present only in nanomolar concentrations endogenously, this high rate suggests a great excess of eIF-2 phosphatase. By contrast, the phosphorylated sites on the β subunit are stable under all conditions studied in lysate (262).

Using [^{32}P]eIF-2α as substrate, an eIF-2 phosphatase has been purified from reticulocyte lysate (263–265). The purified phosphatase has a molecular weight of 95,000 and consists of a 60,000- and a 35,000-dalton subunit (263–265). The enzyme has been identified as a type-2 phosphatase (265). Although phosphorylated eIF-2α is the preferred substrate for the phosphatase, the enzyme will dephosphorylate eIF-2β slowly *in vitro*, in contrast to its lack of effect on eIF-2β in lysate (263, 264).

These characteristics of eIF-2 have assisted in the determination of the number and distribution of phosphorylated sites on eIF-2, since it has permitted the preparation of fully dephosphorylated eIF-2. Highly purified eIF-2 contains two moles of phosphate per mole of eIF-2, by chemical assay of alkali-labile phosphate, and separation of the subunits by carboxymethyl-cellulose chromatography in 8 M urea demonstrates that 95% of the endogenous phosphate is located on the β subunit (263). After dephosphorylation using purified eIF-2 phosphatase, two moles of phosphate may be reintroduced into every mole of β subunit by purified eIF-2β kinase. Only one mole of phosphate per mole of α subunit may be introduced by the eIF-2α specific protein kinase activated in hemin-deficient lysate. Phosphorylation of eIF-2 isolated from hemin-supplemented and hemin-deficient lysate under conditions chosen to preserve the *in situ* phosphorylation state suggests that only 20–30% of the total endogenous eIF-2α pool is phosphorylated in hemin-deficient lysate. The β subunit is found to be maximally phosphorylated in both lysates (263).

5. RELATIONSHIP BETWEEN eIF-2α PHOSPHORYLATION AND PROTEIN SYNTHESIS

For phosphorylation and dephosphorylation of eIF-2 to serve a regulatory role, these functions must in turn be regulated, and relevant signals should evoke appropriate changes in the relative concentrations of the phosphorylated and nonphosphorylated forms. The inhibitory eIF-2α kinase is controlled in an undetermined fashion by hemin. Type 2 phosphoprotein phosphatase activities are regulated not through direct interaction with specific effectors, but by substrate-directed effects in which a metabolite or metal ion combines with the substrate causing it to become a better or worse substrate for the phosphatase (266, 267). The eIF-2 phosphatase conforms to this pattern: its activity is not altered by hemin (262), but is reduced by the change in eIF-2 conformation resulting from its interaction with guanine nucleotides, or the redox state of sensitive sulfhydryl groups (263).

However, since the pool size of eIF-2 is small (268) and only 20–30% of eIF-2 is present in the phosphorylated form at a time when protein synthesis is inhibited by 90%, it is clear that phosphorylation of eIF-2 does not affect its activity in a simple fashion. This value for the extent of phosphorylation is compatible with preliminary data (269).[5] It is also compatible with the effects of sodium selenite, a reversible sulfhydryl reagent (263, 270) that, when added to hemin-supplemented lysate, produces a biphasic pattern of protein synthesis inhibition and an increased phosphorylation of eIF-2α (270). Inhibition can be reversed by the addition of purified eIF-2 to the inhibited lysate. Increasing the concentration of selenite increases the extent of eIF-2α phosphorylation and produces a more rapid onset of translational inhibition; however, the final extent of translational inhibition (>90%) is independent of the extent of eIF-2α phosphorylation (270).

In conjunction with these observations, it should be remembered that one of the criteria to be satisfied in establishing that a particular phosphorylation has regulatory significance is a demonstration that the functional properties of a protein undergo meaningful changes that correlate with the degree of phosphorylation. None of the partial reactions of eIF-2 are affected by *in vitro* phosphorylation (204, 271), except in those laboratories where there appear to be cofactor requirements for the participation of eIF-2 in fractionated, reconstituted sys-

---

[5] Levin *et al.* (269) find less than 50% of the α subunit of eIF-2 present in the phosphorylated form in heme-deficient lysate.

tems and where problems in eIF-2 recovery from the assays have not been eliminated (*124, 126, 127, 140*) (see also Section II,C).

This evidence suggests that phosphorylation of eIF-2 does not result in its inactivation directly, but is preliminary to a second modification that does result in eIF-2 inactivation. Investigations of the characteristics of gel-filtered lysate have uncovered other conditions that affect the activity of eIF-2 in lysate, without alteration of its phosphorylation state, and suggest two additional modifications that affect eIF-2 activity. Impairment of eIF-2 activity is a consequence of gel filtration in the absence of nucleotides, although activation of eIF-2α kinase does not occur (*272, 273*). Similar biphasic kinetics of inhibition are observed as those caused by hemin deprivation unless exogenous eIF-2 is supplied (*272, 273*). Gel filtration in the presence of nucleotides prevents the impairment of eIF-2 activity. The requirement for the presence of nucleotides during gel filtration is not well understood, but suggests that nucleotides are required to maintain eIF-2 in a conformation capable of subsequent catalytic utilization. The gel filtration-induced impairment of eIF-2 activity is overcome by high concentrations of GTP (2 mM), which suggests that the impairment of eIF-2 by gel filtration is caused by interference with its capacity to interact with GTP either directly by conformational changes or indirectly by interference with the catalytic reutilization of (eIF-2) · GDP.

Whether gel filtration is performed in the presence or the absence of nucleotides, the activity of the lysate produced is also dependent on glucose or other NADP-linked substrates (*263, 272–274*). This requirement for NADP-linked substrates reflects the need to maintain critical sulfhydryl groups of eIF-2 in a reduced state, which preserves the factor in an active conformation. Factor eIF-2, reduced by a GSH-regenerating system, has a higher capacity than unmodified eIF-2 to restore protein synthesis in eIF-2-dependent gel-filtered lysate in the absence of glucose. Factor eIF-2, incubated with a glutathione reductase system under conditions favoring the formation of GSH and the maintenance of reduced sulfhydryl groups in proteins, shows a higher capacity than untreated eIF-2 in its ability to restore protein synthesis in the eIF-2-dependent gel-filtered lysate (*273*). Sensitivity of eIF-2 to the redox state of its sulfhydryl groups could provide a means by which protein synthesis is coordinated with the metabolic substrate availability of the cell, as reflected by the redox potential of the NADPH/NADP couple. In the presence of NADP-linked substrates that maintain a high NADPH/NADP ratio, thiol oxidoreductases maintain the unpaired sulfhydryl groups of eIF-2 in a reduced state. If the

NADPH/NADP ratio falls, levels of oxidized glutathione increase and in turn oxidize these unpaired sulfhydryl groups, forming mixed (eIF-2)-glutathione disulfides that are inactive in protein synthesis.

To summarize, two modifications of eIF-2 that result in its inactivity without affecting its phosphorylation state have been identified. The first is the reduced capacity of eIF-2 to interact with GTP, caused by gel filtration. This may be equivalent to a reduced ability of eIF-2 to release bound GDP. The second is the covalent modification of the sulfhydryl groups of eIF-2. Simply stated, phosphorylation of eIF-2, caused by the activation of eIF-2α kinase during hemin deprivation, could either prevent the recycling of (eIF-2) · GDP, or modify the interaction of eIF-2 with oxidation–reduction systems that cycle eIF-2 between a reduced and an oxidized state.

However, it is difficult to differentiate between these alternatives, because many of the effects of GDP on purified eIF-2 are mimicked by oxidation of the sulfhydryl groups of eIF-2. For instance, both high GDP/GTP ratios and eIF-2 sulfhydryl group oxidation reduce ternary complex formation and the accessibility of eIF-2α to eIF-2 phosphatase (263). In addition, there are situations in which the two effects seem to be interrelated. Such a situation is seen in the effects of GTP on the activity of the glucose-dependent gel-filtered lysate, in which glucose is required to maintain critical sulfhydryl groups of eIf-2 in a reduced condition (273). High GTP levels (2 mM) obviate the requirement for NADP-linked substrates in lysate gel-filtered in the presence of nucleotides (273). This suggests that, if changes in the redox state of the sulfhydryl groups of eIF-2 represent the important secondary modification that occurs during hemin deprivation, the effect is exerted at the level of (eIF-2) · GDP recycling. It is of interest that high concentrations of GTP (2 mM) afford protection against the inhibition of protein synthesis by hemin deprivation (241) and reduce the normal utilization of NADPH that occurs during protein synthesis (274).

The possible interrelationships between eIF-2α phosphorylation, the redox state of its sulfhydryl groups, and its capacity to recycle catalytically, may be visualized in two ways as shown in Fig. 8. Scheme 1 in Fig. 8 shows that eIF-2 with reduced sulfhydryl groups is active in ternary complex formation. After hydrolysis of the bound GTP, the (eIF-2 · GDP) complex is inactive, but may be recycled by an unspecified mechanism either by GDP release or by GDP/GTP exchange. After phosphorylation, eIF-2 with reduced sulfhydryl groups is still active in ternary complex formation, but does not recycle efficiently and accumulates as (eIF-2) · GDP, which has a decreased capacity to

Scheme 1

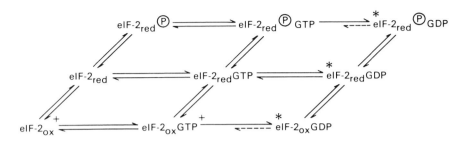

Scheme 2

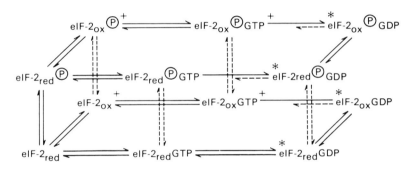

* Inactive
+ May Be Inactive

FIG. 8. Working schemes for the regulation of eIF-2 activity. The (eIF-2) · GDP complexes, in which eIF-2 is either phosphorylated or contains oxidized sulfhydryl groups, are depicted as being unable to regenerate active eIF-2. In the schemes, the conversion of modified (eIF-2) · GDP to modified (eIF-2) · GTP is depicted as the inhibited step. However, it is not known whether this occurs via GDP release or GDP/GTP exchange. In Scheme 1, oxidation of the sulfhydryl groups of eIF-2 or phosphorylation of eIF-2α are seen as unrelated processes, both of which ultimately interfere with the catalytic recycling of eIF-2 leading to the accumulation of inactive (eIF-2) · GDP. In Scheme 2, changes in the phosphorylation and redox states of eIF-2 are seen as interrelated processes.

take part in ternary complex formation. In this scheme, oxidation of the sulfhydryl groups of eIF-2 is seen as an unrelated process that also interferes with the catalytic recycling of eIF-2 leading to the accumulation of inactive (eIF-2) · GDP. Scheme 2 is very similar to Scheme

1, but in Scheme 2 changes in the phosphorylation and redox states of eIF-2 are seen as interrelated processes; phosphorylation favors the oxidation of sulfhydryl groups, and vice versa.

During hemin deprivation, there is a 20–30% chance of eIF-2 becoming inactivated during an initiation cycle, because the steady-state level of phosphorylation is only 20–30%. Consequently, a lag is predicted before protein synthesis becomes inhibited, the length of which would be proportional to eIF-2 pool size and extent of phosphorylation.

6. Reversal Factors

A factor(s) other than eIF-2 relieve(s) the inhibition of protein synthesis caused by hemin deprivation or the hemin-regulated inhibitor (275–280). The activity has been isolated both from postribosomal supernatant and from 500 mM KCl ribosomal wash. From the information presented in early reports, it was not possible to exclude the possibility that the factor was eIF-2 or one of its subunits. However, two recent reports (279, 280) demonstrate that the reversal factor (or antiinhibitor) is a distinct component. Purified and partially characterized reversal factor seems to consist of three subunits with molecular weights of 81,000, 60,000, and 41,000 (279). The factor appears to prevent the inhibition of protein synthesis by a process unrelated to the phosphorylation state of eIF-2α (279).

It is suggested that the factor interacts with and reactivates a form of eIF-2 inactivated by a mechanism distinct from phosphorylation (279). Since it is possible that this factor represents an eIF-2 recycling factor, the determination of its mechanism of action may help to clarify the mechanism of protein synthesis inhibition that occurs during hemin deprivation.

7. Summary of Mechanism of Hemin Control of Translation

In conclusion to this section, a scheme is presented in Fig. 9 outlining the events in initiation affected by hemin deprivation. Some of the factors known to affect eIF-2 activity and their possible interactions with the hemin-mediated control of translation are depicted. It should be pointed out that, although the locus of hemin action has been defined to some extent, it is not known whether its mechanism of action at a molecular level represents an entirely new mechanism for heme action, or whether it is acting in a manner similar to the known functions of heme in electron transport and redox reactions.

Because phosphorylation of eIF-2 provides an inadequate explana-

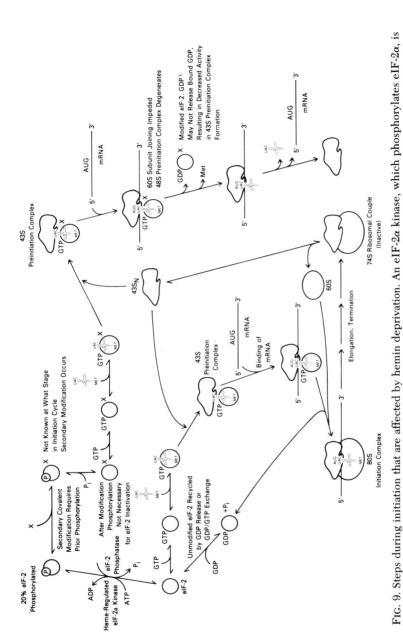

FIG. 9. Steps during initiation that are affected by hemin deprivation. An eIF-2α kinase, which phosphorylates eIF-2α, is activated during hemin deprivation. The phosphate of eIF-2 turns over rapidly owing to the presence of an active eIF-2 phosphatase such that the steady-state level of phosphorylated eIF-2 does not exceed 20 to 30%. In its phosphorylated state, eIF-2 is subject to a secondary modification that is more stable than phosphorylation and is responsible for the inhibition of initiation. It is not known at what stage in initiation the secondary modification occurs. The interaction of 60 S subunit joining is impeded, and the Met-tRNA$_f$ in the defective preinitiation complex is accessible to its hydrolase. After deacylation, the complex becomes unstable and degenerates, giving rise transiently to a >43 S, <48 S complex containing mRNA and tRNA$_f$. Initiation factors −2 and −3 are released. The eIF-2 is released as inactive (eIF-2) · GDP, which does not interact with the eIF-2 recycling mechanism. A pool of inactive (eIF-2) · GDP accumulates at a rate determined by the steady-state phosphorylation of eIF-2, resulting in gradually decreasing rates of 43 S preinitiation complex formation.

tion to account for the observed inhibition of protein synthesis, it is important both to continue the characterization of other covalent modifications of eIF-2 and to determine whether eIF-2 recovered from inhibited lysate is functionally or physically distinct from eIF-2 recovered from lysate actively engaged in protein synthesis. Of outstanding importance is the determination of the normal mechanisms involved in the regeneration of active eIF-2 in the initiation cycle. At present, because of the relative K_D's of GDP and GTP for eIF-2 (K_DGDP = 10^{-8} M, K_DGTP = 10^{-6} M), it is considered probable that there are factors in the lysate that promote either GDP release from eIF-2 or GDP/GTP exchange. The existence of such factors has not been demonstrated, and it is possible that such a factor or factors could be represented by the γ subunit of eIF-2, to which a function has not yet been ascribed.

B. Regulation of Translation by Interferon

1. HISTORICAL INTRODUCTION

The interferon system provides a normal defensive response of animal cells to virus infection, a response that varies with the cell type and virus involved (281). This system represents an example of induced translational control. Although the establishment of the antiviral state by interferon is dependent on nuclear events and requires several hours, its activation by the virus is not, and is caused, in part by an inhibition of translation (282–285). Interferon treatment of cells inhibits the synthesis of viral proteins in cells, suggesting a block in translation (282, 285). This suggestion was reinforced by the similarity between the effects of conventional inhibitors of protein synthesis and interferon treatment of virus-infected cells (283).

To approach the molecular basis of interferon action, cell-free systems were developed that display interferon-mediated inhibition of translation. Initially, cell-free systems were prepared from virus-infected cells in which the translation of viral mRNA is inhibited by treatment of the cells with interferon (286–290). Kerr and co-workers (291) found that an inhibitory effect on initiation predominates over an inhibition of elongation. This observation is supported by studies of intact vaccinia-infected L cells (285). Mixing experiments, using the cell-free systems, indicated that reduced initiation of protein synthesis arises from a dominant inhibitor associated with both the ribosomal fraction and the postribosomal supernatant (287–290). Formation of this inhibitor, which is not interferon itself, correlates well with the development of the antiviral state (288). In intact cells, both processes require several hours to develop, and both are blocked by actinomycin D (288). Translation of cellular mRNA is inhibited to the same extent as viral mRNAs as a consequence of viral infection in extracts from interferon-treated cells (288, 289, 292). Further simplification of the interferon-sensitive, cell-free, protein-synthesizing system was made possible by the discovery that infection prior to the preparation of cell extracts can be omitted by the addition of double-stranded DNA to the *in vitro* extract (293). These investigations demonstrated that translation of encephalomyelocarditis virus RNA in cell-free extracts of interferon-treated L cells is inhibited in the presence of double-stranded RNA (293). Extracts prepared from L cells not treated with interferon show little sensitivity to double-stranded RNA in its ability to translate EMC RNA (293).

Translational inhibitors, first observed in interferon-treated, virus-infected cell extracts, are also present in the ribosomal wash and postribosomal supernatant of extracts prepared from interferon-treated cells after incubation with double-stranded RNA (293). The action of inhibitors is dependent on ATP, as well as dsRNA (294–296). One inhibitor exibits a double-stranded RNA-dependent protein kinase activity, reminiscent of the inhibitory activity found in the postribosomal supernatant of rabbit reticulocyte lysate incubated in the absence of hemin (245). However, in addition to the kinase, a heat-stable, low-molecular-weight inhibitor of protein synthesis (297), as well as several enzyme activities, are found in dsRNA-treated extracts from interferon-treated cells (298).

The interrelationships of these enzyme activities and their relevance to the mechanism by which interferon blocks viral replication *in vivo* are not clear. However, considerable information is available on the characteristics of the components studied *in vitro*.

2. DOUBLE-STRANDED RNA-DEPENDENT PROTEIN KINASE(S)

When extracts of interferon-treated cells are incubated in the presence of dsRNA, at least two proteins become phosphorylated; one of molecular weight 67,000 is thought to be the inhibitory kinase (299); the other, of molecular weight 35,000, is its substrate, eIF-2α (296, 300–303). In rabbit reticulocyte lysate incubated with dsRNA, phosphorylation of the same proteins are observed (245, 251, 304).

Documentation of the relationship between the induction of protein kinase activity and the previous exposure of the cells to interferon is the observation that there is a time-dependent induction of a kinase susceptible to activation by dsRNA (298, 305). This activity is stimulated from a measurable basal level up to fivefold greater activity in L cells by increasing interferon concentrations (306–309) and is blocked by actinomycin D (306). Rabbit reticulocytes contain relatively high levels of dsRNA-dependent kinase activity without interferon pretreatment (245, 251, 304). The dsRNA-dependent kinase activity copurifies with the 67,000-dalton protein and is a potent inhibitor of both endogenous and virus RNA-directed translation in nucleated cell extracts, as well as endogenous protein synthesis in reticulocyte lysate (301, 310, 311). The inhibition is presumed to operate by phosphorylation of eIF-2α in a fashion similar to the inhibition of reticulocyte lysate by the heme-regulated inhibitory kinase. The two mechanisms are subject to the same speculations (see Section II,A,5). Once activated by dsRNA, the kinase remains active, even after degradation of dsRNA by RNase III (311).

The kinase can be purified approximately 1000-fold by single-step chromatography on poly(I · C)-Sepharose (299). However, this provides a heterogeneous preparation containing distinct and separable histone kinases (299). Although certain purification procedures appear to result in a diminution in the ability of dsRNA to stimulate the kinase (312), there is reported a 90–95% pure preparation of the kinase that retains its dsRNA-dependency (313). The dsRNA-dependent kinase for eIF-2α is cAMP-independent, and migrates as a 67,000-dalton polypeptide on dodecyl sulfate/polyacrylamide gels (312). Activation of the kinase appears to involve autophosphorylation, although the possibility of a kinase phosphorylating enzyme as a minor contaminant has not been excluded (313). On the basis of its ribosomal localization (245, 304), its size (245, 299, 313, 314), its substrate specificity (302, 309, 311), and its antigenicity (315), the dsRNA-dependent kinase is distinct from the heme-regulated inhibitory kinase. Nevertheless, the dsRNA-dependent kinase catalyzes the phosphorylation of sites on eIF-2α indistinguishable from those phosphorylated by the heme-regulated kinase as discerned by thin-layer chromatographic analysis of ^{32}P-phosphorylated tryptic digests (316). However, these fragments were not subjected to amino-acid sequence analysis, required to establish their identity unequivocally.

There is, as yet, no evidence that a dsRNA-dependent kinase is activated in intact interferon-treated, virus-infected cells, although there is eIF-2α kinase activity in intact L cells incubated with dsRNA (317).

3. (2′-5′)-Oligoadenylate and (2′-5′)-Oligoadenylate Synthetase

Roberts et al. (297) reported the formation of a low-molecular-weight inhibitor of protein synthesis in extracts of interferon-treated L cells incubated with dsRNA and ATP. Subsequent studies led to the isolation of this inhibitor and its characterization as (2'-5')-oligoadenylate *(318–320)*. The major component encountered is the trinucleotide, pppA(2'-5')A(2'-5')A *(320)*. Other oligonucleotides with a similar structure may be present; the general formula for the series is pppA(2'-5'A)$_n$ ($n = 1-10$) *(320)*. All oligonucleotides larger than the dimer inhibit protein synthesis. Double-stranded RNA-dependent synthesis of (2'-5')A$_n$ also occurs in extracts of interferon-treated chick embryo *(321)*, 3T6 *(307)*, HeLa *(322)*, and Ehrlich ascites cells *(312, 323)*, as well as in rabbit reticulocyte lysate in the absence of interferon pretreatment *(324, 325)*.

Upon exposure of cells to interferon, dsRNA-dependent (2'-5')A$_n$-synthetase activity increases linearly for several hours before reaching a steady state *(326, 327)*. The increase in dsRNA-dependent activity varies with different cell lines and has been found to be as high as 10,000-fold in interferon-treated chick embryo cells *(327)*. *De novo* synthesis of the enzyme is involved, which is dependent on the dose and time of exposure of the cells with interferon and is blocked by actinomycin D *(327)*.

Partial purifications of the synthetase have been reported *(309, 323)*, and a 56,000 dalton polypeptide, which appears in chick embryo cells after interferon treatment and co-chromatographs with (2'-5')A$_n$-synthetase activity, has been observed *(327)*. In addition, after an initial purification on poly(I · C)-Sepharose, the (2'-5')A$_n$-synthetase has been separated from the dsRNA-dependent kinase activity on DEAE cellulose *(299)*. However, the resultant "kinase-free" synthetase required a second, as yet unidentified, component in addition to dsRNA, for optimal activity *(299)*. An apparently homogeneous preparation of (2'-5')A$_n$-synthetase that retains its dsRNA-dependency has been made; it has a native molecular weight of 85,000 and, by dodecyl sulfate electrophoresis, a molecular weight of 105,000 *(328)*. The relationship between this protein and the 56,000-dalton polypeptide *(327)* is at present unclear.

The (2'-5')-oligoadenylate inhibits both endogenous and EMC RNA-directed protein synthesis in extracts from interferon-treated L cells and reticulocyte lysate *(318, 329)*. The kinetics of inhibition in these systems are similar; protein synthesis proceeds normally for at least 15 minutes and then continues only at a much reduced rate *(318, 329)*. If reticulocyte lysates or L-cell extracts are incubated with the oligonucleotide before the addition of ATP, GTP, and amino acids to the mixture, the lag period is considerably reduced *(329)*. This distinguishes these biphasic kinetics of inhibition from those caused by hemin deprivation in rabbit reticulocyte lysate, and is suggestive of a different mechanism of action.

4. (2'-5')-Oligoadenylate-Activated Endonuclease Activity

Much evidence indicates that an endonuclease is activated by $(2'\text{-}5')A_n$ at subnanomolar concentrations (329–331), and that the inhibition of protein synthesis by $(2'\text{-}5')A_n$ in cell-free extracts is the result of endonuclease activation and subsequent degradation of mRNA (329, 330). Incubation of extracts from interferon-treated cells with dsRNA results in mRNA cleavage, suggestive of the activation of an endonuclease (300, 331, 332). Also, the inhibition of protein synthesis by $(2'\text{-}5')A_n$ is accompanied by polysome breakdown, caused not by an inhibition of initiation of protein synthesis, but by degradation of mRNA (329, 330). The inhibition of protein synthesis by $(2'\text{-}5')A_n$ cannot be reversed by eIF-2, but can be overcome by the addition of mRNA (329).

The $(2'\text{-}5')A_n$-activated endonuclease, of molecular weight 185,000 (332), is present in both interferon-treated and untreated cells and also in reticulocyte lysate (331). Its activity is dependent on the continuous presence of $(2'\text{-}5')A_n$; when the activator is removed, the enzyme reverts to an inactive state (331). Activation is not accompanied by a major size change of the native protein (332a). It is suggested that, *in vivo*, the enzyme could be activated in a transient way if oligonucleotides are formed in limited amounts and rapidly degraded (298). The $(2'\text{-}5')A_n$ will inhibit protein synthesis in intact, but permeabilized, BHK cells, and such inhibition is caused by the transient activation of a nuclease (333).

5. (2'-5')-Phosphodiesterase Activity

Following the addition of $(2'\text{-}5')A_n$ to L-cell extracts, there is a rapid loss of the oligonucleotide, paralleled by a decline in the induced endonucleolytic activity. The rate of degradation is similar in extracts prepared from control or interferon-treated cells (331). Degradation results from the action of an unidentified (2'-5')-phosphodiesterase activity. It remains to be shown whether $(2'\text{-}5')A_n$ undergoes rapid catabolism in whole cells.

6. Relative Importance of Double-Stranded RNA-Dependent Enzyme Activity

Although it is possible to assess the relative importance of the different interferon-induced dsRNA-dependent enzymes in cell-free systems like reticulocyte lysate, it is more difficult to assess the interaction of the two apparently unrelated dsRNA-mediated mechanisms of inhibition in intact virus-infected cells. In lysates of rabbit reticulocytes, for instance, the dsRNA-activated inhibitory kinase responds to a lower range of dsRNA concentrations (10^{-9} to 10^{-7} g/ml) than does $(2'-5')A_n$-synthetase ($>10^{-7}$ g/ml) (325). Also, the activation of the inhibitory kinase occurs more rapidly than the production of $(2'-5')A_n$, and the former is capable of inhibiting protein synthesis at an earlier stage. A consideration of the kinetics of inhibition (325), coupled with a knowledge that the site of the lesion is predominantly at 43 S preinitiation complex formation (301), which is partially reversible by addition of eIF-2, suggests that at dsRNA concentrations from 10^{-9} to 10^{-7} g/ml, the initial inhibition is mediated by the inhibitory kinase. However, the translation of both endogenous mRNA and viral mRNA is inhibited in dsRNA-treated extracts of interferon-treated cells (288, 289, 292).

This contrasts with the characteristics of some interferon-treated intact cells. For instance, during SV40 infection of interferon-treated L cells, total protein synthesis is not affected (334, 335). Similarly, during reovirus infection, cellular protein synthesis is not shut off rapidly, although the accumulation of viral proteins is diminished in interferon-treated cells (335). This apparent discrimination between host and viral mRNA does not seem consistent with a mechanism involving the inhibitory kinase. However, a possible mechanism giving rise to discrimination is provided by the $(2'-5')A_n$-polymerase/endonuclease system. The average half-life of reovirus mRNA is approximately 12.7 hours in control cells, but only 4.8 hours in interferon-treated cells (336). It is postulated that the $(2'-5')A_n$ synthetase is activated by partially double-stranded replicative intermediates of RNA viruses, and that the $(2'-5')A_n$ synthesized may activate its synthetase locally (327); viral mRNA would then be cleaved preferentially. Supporting this hypothesis is the demonstration that mRNA covalently linked to dsRNA in extracts of interferon-treated HeLa cells is preferentially degraded compared with mRNA not linked to dsRNA (337).

A suggestion has been made that the $(2'-5')A_n$ synthetase may play a role in normal cell function in the control of proliferation and development (338). Stark and co-workers (338) reported a wide distribution of $(2'-5')A_n$-synthetase activity in mammalian cells, including reticulocytes, human lymphoblastoid cells, mouse erythroleukemia cells, and chick oviduct, after withdrawal of estrogen. It has also been observed in lymphocytes from normal mouse spleen and in other lymphoid tissues including thymocytes and splenocytes (339). Also, Revel (340) has reported higher $(2'-5')A_n$ levels in concanavalin-stimulated mouse splenic lymphocytes. In contrast to the effect in interferon-treated cells, this increase seems to be related to an increase in $(2'-5')$-phosphodiesterase activity in Con A-stimulated lymphocytes (340).

IV. Summary: Overview on Present Understanding of the Control of Initiation

The regulation of the rate of initiation has been shown to be of great importance in many cells and tissues during rapid transitions from one metabolic state to another. Although mechanisms by which initiation is regulated in reticulocyte lysate and in extracts from interferon-treated cells have been determined, it is not known to what extent these mechanisms represent general regulatory mechanisms that may be found in a wide variety of cell types and physiological conditions, such as those discussed in Section I.

During amino-acid or glucose starvation in Krebs II ascites cells, a drop in the rate of initiation correlates with a decrease in the level of 43 S preinitiation complexes (*341*). This suggests that initiation in these cells may be controlled by changes in the activity of eIF-2 by mechanisms similar to those regulating eIF-2 activity in reticulocyte lysate by hemin and in interferon-treated cell-free extracts by double-stranded RNA. However, no definitive answers are yet available for a variety of reasons. It has not yet been possible to prepare cell-free protein-synthesizing systems from nucleated cells that exhibit comparable activity with the intact cells. Cell-free extracts prepared from nucleated cells exhibit a very low capacity to take part in initiation. Many cell-free extracts described from nucleated cells do not exhibit any reinitiation; the only protein synthetic activity observed represents completion of peptides on polysomes formed in the intact cell (*342*). Cell-free extracts from nucleated cells that exhibit a capacity to reinitiate have been described, but the best of these give only two to three cycles per ribosome in 15–20 minutes, with initiation declining with incubation time (*343, 344*). This contrasts markedly with the reticulocyte lysate in which there occurs almost one initiation cycle per ribosome per minute in the presence of hemin. Even in the absence of hemin, each ribosome reinitiates four or five times. As a consequence of this, experiments of the type performed in reticulocyte to investigate the underlying mechanisms that result in a decrease in initiation are not possible in other model systems.

The reasons underlying the failure to produce active cell-free systems are not understood, but may be connected, in part, with higher levels of ribonuclease and proteolytic enzyme activity. However, it is possible that the reticulocyte represents a slightly simpler protein synthetic system. This is suggested by the different effects of histidinol, a histidine analog, in reticulocytes and reticulocyte lysate compared with many nucleated cells. In Krebs ascites cells, incubation with histidinol results in a decrease in initiation (*345*). In reticulocytes (346, 346a) and reticulocyte lysate (89), histidinol only inhibits elongation. It is possible that reticulocytes lack a regulatory component of initiation present in other cell types.

Without the advantages of cell-free protein-synthesizing systems, experiments of the type performed in reticulocyte lysates to investigate the underlying mechanisms regulating initiation in the nucleated cell model systems have not been possible. Krebs ascites cells starved for amino acids or glucose contain an inhibitor of reticulocyte lysate protein synthesis, the effects of which may be reversed by addition of eIF-2 (347). However, it is not known whether the inhibition caused by this Krebs ascites cell extract in reticulocyte lysate also results in phosphorylation of reticulocyte eIF-2α. It has not been possible to demonstrate a change in the phosphorylation state of eIF-2α in Krebs ascites cells starved for glucose or amino acids, although precautions against phosphatase activity during isolation have not been taken. There are reports of inhibitors of initiation from many nucleated cells (60, 348–351), the best characterized of which is that isolated from rat liver (349). From its effects on protein synthesis in reticulocyte lysates, its chromatographic behavior, and its ability to phosphorylate eIF-2α, the rat liver inhibitor resembles eIF-2α kinase. However, as with all the other inhibitors described, its physiological significance and *in vivo* regulation are unknown.

In summary, the accumulation of information on both the sequence of events and the regulation of initiation derived from reticulocyte lysate has provided a solid foundation from which to ask questions of a wide range of cells and tissues. However, it is not yet known if a ubiquitous regulatory mechanism exists for the control of initiation, or if the regulation of eIF-2 activity is the prevalent point of control. Although data are accumulating to suggest that the control of eIF-2 activity may represent a general regulatory mechanism, it is becoming increasingly apparent that more than one step in the initiation pathway is subject to control. It is expected that the development of cell-free protein synthesizing systems from nucleated cells will allow the pursuit of these questions.

Acknowledgments

We wish to thank Drs. M. C. Clemens, P. J. Farrell, E. C. Henshaw, V. M. Pain, H. Trachsel, and H. O. Voorma for providing us with manuscripts prior to publication. We are grateful also to Drs. D. M. F. Cooper and W. C. Merrick for interesting discussions. Thanks are expressed to Bill Mapes for skillful technical illustrations, and to Kirsten Cook and Joy Grant for expert editorial assistance. We are grateful to Eve Church for her patient proofreading.

REFERENCES

1. W. F. Anderson, L. Bosch, W. E. Cohn, H. Lodish, W. C. Merrick, H. Weissbach, H. G. Wittmann and I. G. Wool, *FEBS Lett.* **76**, 1 (1977).
1a. M. Grunberg-Manago and F. Gros, This Series **20**, 209 (1977).
1b. H. F. Lodish, *Nature* **215**, 385 (1974).
2. H. F. Lodish, *ARB* **45**, 39 (1976).
3. J. E. Bergmann and H. F. Lodish, *JBC* **254**, 11927 (1979).
4. B. E. H. Maden, M. H. Vaughan, J. R. Warner and J. E. Darnell, *JMB* **45**, 265 (1969).
5. M. T. Franze-Fernandez and A. O. Pogo, *PNAS* **68**, 3040 (1971).
6. M. T. Franze-Fernandez and A. V. Fontanive-Senguesa, *BBA* **331**, 71 (1973).
7. J. R. Warner, *in* "Ribosomes" (M. Numura, A. Tissières, and P. Lengyel, eds.), p. 461. Cold Spring Harbor Lab., Cold Spring Harbor, New York, 1976.
8. D. J. Millward, P. J. Garlick, R. J. C. Stewart, D. O. Nnamyelugo and J. C. Waterlow, *BJ* **150**, 235 (1975).
9. H. N. Munro, *in* "Mammalian Protein Metabolism" (H. N. Munro and J. B. Allison eds.), Vol. 1, p. 381. Academic Press, New York, 1964.
10. E. C. Henshaw, *in* "Biochemistry of Cellular Regulation" (M. Clemens, ed.), Chap. 5, p. 119. CRC Press, West Palm Beach, Florida, 1980.
11. V. R. Young and S. D. Alexis *J. Nutr.* **96**, 255 (1968).
12. V. R. Young, S. C. Stothers and G. Vilaire, *J. Nutr.* **101**, 1379 (1971).
13. E. Moskin, S. Kimata and J. J. Skillman, *Circ. Res.* **30**, 690 (1972).
14. H. E. Morgan, A. C. Hjalmarson and D. E. Rannels, *Recent Adv. Stud. Card. Struct. Metab.* **3**, 561 (1972).
15. R. L. Kao, D. E. Rannels, V. Whitman and H. E. Morgan, *Recent Adv. Stud. Card. Struct. Metab.* **12**, 105 (1976).
16. A. W. Everett, R. R. Taylor and M. P. Sparrow, *Recent Adv. Stud. Card. Struct. Metab.* **12**, 35 (1976).
17. A. Fleck, J. Shepherd and H. N. Munro, *Science* **150**, 628 (1965).
18. T. E. Webb, G. Blobel and V. R. Potter, *Cancer Res.* **26**, 253 (1966).
19. S. H. Wilson and M. B. Hoagland, *BJ* **103**, 556 (1967).
20. E. C. Henshaw, C. A. Hirsh, B. E. Morton and H. H. Hiatt, *JBC* **246**, 436 (1971).
21. K. Tsukada, T. Moriuama, T. Umeda and I. Lieberman, *JBC* **243**, 1160 (1968).
22. H. A. Johnson and J. M. V. Roman, *Am. J. Pathol.* **49**, 1 (1967).
23. R. Kao, D. E. Rannels and H. E. Morgan, *Circ. Res., Suppl.* **38**, I-124 (1972).
24. M. Ayuso-Parilla and R. Parilla, *EJB* **55**, 593 (1975).
25. V. M. Pain and P. J. Garlick, *JBC* **249**, 4510 (1974).
26. D. E. Rannels, L. S. Jefferson, A. C. Hjalmarson, E. B. Wolpert and H. E. Morgan, *BBRC* **40**, 1110 (1970).
27. H. E. Morgan, L. S. Jefferson, E. B. Wolpert and D. E. Rannels, *JBC* **246**, 2163 (1971).
28. P. M. Sender and P. J. Garlick, *BJ* **132**, 603 (1973).
29. L. S. Jefferson, J. O. Koehler and H. E. Morgan, *PNAS* **69**, 816 (1972).
30. L. S. Jefferson, J. B. Li and S. R. Rannels, *JBC* **252**, 1476, (1976).
31. A. H. Korner, *Ann. N. Y. Acad. Sci.* **148**, 408 (1968).
32. A. C. Hjalmarson, D. E. Rannels, R, Kao and H. Morgan, *JBC* **250**, 4556 (1975).
33. E. Kaminskas, *JBC* **247**, 5470 (1972).
34. E. Kaminskas, *J. Cell. Physiol.* **82**, 475 (1973).
35. E. Kaminskas, *Exp. Cell Res.* **94**, 7 (1975).

36. W. J. W. VanVenrooij, E. C. Henshaw and C. A. Hirsch, *JBC* **245**, 5947 (1970).
37. W. J. W. Vanvenrooij, E. C. Henshaw and C. A. Hirsch, *BBA* **259**, 127 (1972).
38. M. H. Vaughan, P. J. Pawlowski and J. Forchhammer, *PNAS* **58**, 2057 (1971).
39. S. Y. Lee, V. Krsmanovic and G. Brawerman, *Bchem* **10**, 895 (1971).
40. E. Eliasson, G. E. Bauer and T. Hultin, *J. Cell Biol.* **33**, 287 (1967).
41. S. L. Mendelsohn, S. K. Nordeen and D. A. Young, *BBRC* **79**, 53 (1977).
42. N. R. Ling and J. E. Kay, *in* "Lymphocyte Stimulation" (N. R. Ling and J. E. Kay, eds.), 2nd ed., Chapter 13, p. 324. North-Holland Publ., Amsterdam, 1975.
43. B. Westermark and Å. Wasterson, *Adv. Metab. Disord.* **8**, 85 (1975).
44. R. Ross and A. Vogel, *Cell* **14**, 203 (1978)
45. C. H. Heldin, B. Westermark and A. Wasterson, *PNAS* **76**, 3722 (1979).
46. P. Felig and J. Wahren, *FP* **33**, 1092 (1974).
47. P. Felig, *ARB* **44**, 933 (1975).
48. M. J. Clemens and V. M. Pain, *BBA* **361**, 345 (1974).
49. G. E. Sonenshein and G. Brawerman, *EJB* **73**, 307 (1977).
50. C. P. Stanners and H. Becker, *J. Cell. Physiol.* **77**, 31 (1971).
51. G. J. Todaro, G. K. Lozar and H. Green, *J. Cell. Physiol.* **66**, 325 (1965).
52. E. M. Levine, Y. Becker, C. W. Boone and H. Eagle, *PNAS* **53**, 350 (1965).
53. G. A. Wark and P. G. W. Plageman, *J. Cell. Physiol.* **73**, 213 (1969).
54. H. Becker, C. P. Stanners and J. E. Kudlow, *J. Cell. Physiol.* **77**, 43 (1971).
55. P. S. Rudland, *PNAS* **71**, 750 (1973).
56. R. Jagus and J. E. Kay, *EJB* **100**, 503 (1979).
57. J. E. Kay, T. Ahern and M. Atkins, *BBA* **247**, 332 (1971).
58. T. Ahern and J. E. Kay, *Exp. Cell Res.* **92**, 513 (1975).
59. H. L. Cooper and R. Brawerman, *J. Cell. Physiol.* **93**, 213 (1977).
60. J. E. Kay, D. M. Wallace, C. R. Benzie and R. Jagus, *in* "Cell Biology and Immunology of Leukocyte Function" (M. R. Quastel, ed.), p. 107. Academic Press, New York, 1979.
61. P. S. Rudland, *PNAS* **71**, 750 (1974).
62. I. Emanilov, D. D. Sabatini, J. A. Lake and C. Freienstein, *PNAS* **75**, 1389 (1978).
63. B. Safer, W. F. Anderson and W. C. Merrick, *JBC* **250**, 9067 (1975).
64. B. Safer, S. L. Adams, W. M. Kemper, K. W. Berry, M. Lloyd and W. C. Merrick, *PNAS* **73**, 2584 (1976).
65. P. M. Prichard and W. F. Anderson, *in* "Methods in Enzymology" (L. Grossman and K. Moldave, eds.), Vol. 30, p. 136. Academic Press, New York, 1974.
66. W. M. Kemper, K. W. Berry and W. C. Merrick, *JBC* **251**, 5551 (1976).
67. W. C. Merrick, W. M. Kemper and W. F. Anderson, *JBC* **250**, 5556 (1975).
68. W. C. Merrick, *in* "Methods in Enzymology" (B. W. O'Malley and J. G. Hardman, eds.), Vol. 40, Part E, p. 101. Academic Press, New York, 1975.
69. M. H. Schreier and T. Staehelin, *Nature NB* **242**, 35 (1973).
70. M. H. Schreier and T. Staehelin, *in* "Regulation of Transcription and Translation in Eukaryotes" (E. K. F. Bautz, P. Karlson, and H. Kersten, eds.), p. 335. Springer-Verlag, Berlin and New York, 1973.
71. M. H. Schreier, B. Erni and T. Staehelin, *JMB* **116**, 727 (1977).
72. T. Staehelin, B. Erni and H. Schreier, *in* "Methods in Enzymology" (B. W. O'Malley and J. G. Hardman, eds.), Vol. 40, Part E, p. 136. Academic Press, New York, 1975.
73. R. Benne and J. W. B. Hershey, *PNAS* **73**, 3005 (1976).
74. G. A. Floyd, W. C. Merrick and J. A. Traugh, *EJB* **96**, 277 (1979).
75. R. Benne, C. Wong, M. Luedi and J. W. B. Hershey, *JBC* **251**, 7675 (1976).

76. R. Benne, M. Luedi and J. W. B. Hershey, *JBC* **252**, 5798 (1977).
77. R. Benne, M. Brown-Luedi and J. W. B. Hershey, *JBC* **253**, 3070 (1978).
78. R. Benne, M. Brown-Luedi and J. W. B. Hershey, *in* "Methods in Enzymology" (B. W. O'Malley and J. G. Hardman, eds.), Vol. 40, Part E, p. 15. Academic Press, New York, 1975.
79. H. O. Voorma, A. Thomas, H. Goumans, H. Amesz and C. vander Mast, *in* "Methods in Enzymology" (B. W. O'Malley and J. G. Hardman, eds.), Vol. 40, Part E, p. 124. Academic Press, New York, 1975.
80. H. Trachsel, B. Erni, M. H. Schreier, L. Braun and T. Staehelin, *BBA* **561**, 484 (1979).
81. H. Sonenberg, M. A. Morgan, W. C. Merrick and A. J. Shatkin, *PNAS* **45**, 4843 (1978).
82. J. E. Bergmann, H. Trachsel, N. Sonenberg, A. J. Shatkin and H. F. Lodish, *JBC* **254**, 1440 (1979).
83. N. Sonenberg, K. M. Rupprecht, S. M. Hecht and A. J. Shatkin, *PNAS* **76**, 4345 (1979).
84. H. Trachsel, N. Sonenberg, A. J. Shatkin, J. K. Rose, K. Leong, J. E. Bergman, J. Gordon and D. Baltimore, *PNAS* **77**, 770 (1980).
85. R. Kaempfer, *in* "Ribosomes" (M. Nomura, A. Tissières, and P. Lengyel, eds.), p. 679. Cold Spring Harbor Lab., Cold Spring Harbor, New York, 1974.
86. A. K. Falvey and T. Staehelin, *JMB* **53**, 21 (1970).
87. B. Safer, W. Kemper and R. Jagus, *JBC* **253**, 3384 (1978).
88. B. Safer, R. Jagus and W. Kemper, *in* "Methods in Enzymology" (B. W. O'Malley and J. G. Hardman, eds.), Vol. 40, Part E, p. 61. Academic Press, New York, 1975.
89. R. Jagus and B. Safer, *JBC* **254**, 6865 (1979).
90. R. Kaempfer and M. Meselson, *CSHSQB* **34**, 209 (1969).
91. S. Ottolenghi, P. Comi, B. Gialioni, A. M. Ganni and G. G. Guidotti, *EJB* **33**, 227 (1973).
92. E. Henshaw, D. Guiney and C. Hirsch, *JBC* **248**, 4367 (1963).
93. R. A. Mathew and F. O. Wettstein, *BBA* **366**, 300 (1974).
94. H. A. Thompson, I. Sadnik, J. Scheinbuks and K. Moldave, *Bchem* **16**, 2221 (1977).
95. S. M. Heywood, D. S. Kennedy and A. J. Bester, *PNAS* **71**, 2428 (1974).
96. D. Peterson, W. C. Merrick and B. Safer, *JBC* **254**, 2509 (1979).
97. R. Benne and J. W. B. Hershey, *JBC* **253**, 3078 (1978).
98. R. Kaempfer and J. Kaufman, *PNAS* **69**, 3317 (1972).
99. B. Safer, D. Peterson and W. C. Merrick, *in* "Translation of Synthetic and Natural Polynucleotides" (A. B. Legocki, ed.), p. 24. University of Agriculture, Poznan, Poland, 1977.
100. A. Thomas, H. Houmans, H. O. Voorma and R. Benne, *EJB* **107**, 39 (1980).
101. J. Morrisey and B. Hardesty, *ABB* **152**, 385 (1972).
102. N. K. Gupta and K. Aerni, *BBRC* **51**, 907 (1973).
103. J. M. Cimadevilla, J. Morrisey and B. Hardesty, *JMB* **83**, 437 (1974).
104. J. McCuiston, R. Parker and K. Moldave, *ABB* **172**, 387 (1976).
105. O. Nygård and T. Hultin, *EJB* **72**, 537 (1977).
106. K. B. Andersen, *EJB* **96**, 109 (1979).
107. B. Safer, S. L. Adams, W. F. Anderson and W. C. Merrick, *JBC* **250**, 9076 (1975).
108. D. H. Levin, D. Kyner and G. Acs, *PNAS* **70**, 41 (1973).
109. N. K. Gupta, C. L. Woodley, Y. C. Chen and K. K. Bose, *JBC* **248**, 4500 (1973).
110. G. L. Dettman and W. M. Stanley, Jr., *BBA* **299**, 142 (1973).
111. R. S. Ranu and I. G. Wool, *JBC* **251**, 1926 (1976).

112. A. Barrieux and M. G. Rosenfeld, *JBC* **252**, 392 (1977).
113. M. A. Lloyd, J. C. Osborne, B. Safer, G. Powell and W. C. Merrick, *JBC* **255**, 1189 (1980).
114. A. Barrieux and M. G. Rosenfeld, *JBC* **252**, 3843 (1977).
115. E. A. Stringer, A. Chaudhuri and U. Maitra, *JBC* **254**, 6845 (1979).
116. G. L. Dettman and W. M. Stanley, Jr., *BBA* **287**, 124 (1972).
117. Y. C. Chen, C. L. Woodley, K. K. Bose and N. K. Gupta, *BBRC* **48**, 1 (1972).
118. G. M. Walton and G. N. Gill, *BBA* **390**, 231 (1975).
119. G. M. Walton and G. N. Gill, *BBA* **447**, 11 (1976).
120. E. A. Stringer, A. Chaudhuri and U. Maitra, *BBRC* **76**, 586 (1977).
121. B. Safer and R. Jagus, *PNAS* **76**, 1094 (1979).
122. Y. C. Chen, C. L. Woodley, K. K. Bose and N. R. Gupta, *BBRC* **46**, 839 (1972).
123. A. Barrieux and M. G. Rosenfeld, *JBC* **252**, 1843 (1977).
124. A. Dasgupta, A. Majumdar, A. D. George and N. K. Gupta, *BBRC* **71**, 1234 (1976).
125. A. Dasgupta, A. Das, R. Roy, R. Ralston, A. Majumdar and N. K. Gupta, *JBC* **253**, 6054 (1978).
126. A. Das and N. K. Gupta, *BBRC* **78**, 1433 (1977).
127. C. deHaro, A. Datta and S. Ochoa, *PNAS* **75**, 243 (1978).
128. C. deHaro and S. Ochoa, *PNAS* **75**, 2713 (1978).
129. C. deHaro and S. Ochoa, *PNAS* **76**, 1741 (1978).
130. C. deHaro and S. Ochoa, *PNAS* **76**, 2163 (1978).
131. R. S. Ranu, I. M. London, A. Das, A. Dasgupta, A. Majumdar, R. Ralston and N. K. Gupta, *PNAS* **75**, 745 (1978).
132. R. S. Ranu and I. M. London, *PNAS* **76**, 1079 (1979).
133. B. Benne, H. Amesz, J. W. B. Hershey and H. O. Voorma, *JBC* **254**, 3201 (1979).
134. H. Trachsel, B. Erni, M. H. Schreier and T. Staehelin, *JMB* **116**, 755 (1977).
135. K. E. Smith and E. C. Henshaw, *Bchem* **14**, 1060 (1975).
136. D. Levin, D. Kynes and G. Acs, *JBC* **248**, 6416 (1973).
137. A. Das, R. O. Ralston, M. Grace, R. Roy, P. Ghosh-Dastridas, H. K. Das, B. Yaghmai, S. Palmer and N. K. Gupta, *PNAS* **76**, 5076 (1979).
138. B. Erni, E. T. H. Dissertation No. 5862. Basel Institute for Immunology, Basel (1976).
139. I. C. Sundkvist and T. Staehelin, *JMB* **99**, 401 (1978).
140. A. Majumdar, R. Roy, A. Das, A. Dasgupta and N. K. Gupta, *BBRC* **78**, 161 (1977).
141. R. O. Ralston, A. Das, M. Grace, H. Das and N. K. Gupta, *PNAS* **76**, 5076 (1979).
142. N. K. Gupta, B. Chatterjee and A. Majumdar, *BBRC* **65**, 797 (1975).
143. O. W. Odom, G. Kramer, A. B. Henderson, P. Pinphanichakorn and B. Hardesty, *JBC* **253**, 1807 (1978).
143a. H. Trachsel and T. Staehelin, *PNAS* **75**, 204 (1978).
144. A. Van der Voorde, R. Contreras, R. Rogers and W. Fiers, *Cell* **9**, 117 (1976).
145. D. T. Wigle and A. Smith, *Nature NB* **242**, 136 (1973).
146. F. Rottman, A. J. Shatkin and R. P. Perry, *Cell* **3**, 197 (1974).
147. A. J. Shatkin, *Cell* **9**, 645 (1976).
148. W. Filipowicz, *FEBS Lett.* **96**, 1 (1978).
149. M. Kozak and A. J. Shatkin, *JMB* **112**, 75 (1977).
150. R. Leibowitz and S. Penman, *J. Virol.* **8**, 661 (1971).
151. Y. Kaufman, E. Goldstein and S. Penman, *PNAS* **73**, 1834 (1976).
152. J. K. Rose, H. Trachsel, K. Leon and D. Baltimore, *PNAS* **75**, 2732 (1978).
153. T. Helentjaris and E. Ehrenfeld, *J. Virol.* **26**, 510 (1978).

154. D. A. Shafritz, J. A. Weinstein, B. Safer, W. C. Merrick, L. A. Weber, E. D. Hickey and C. Baglioni, *Nature* **261**, 291 (1976).
155. N. Sorenberg and A. J. Shatkin, *PNAS* **74**, 4288 (1977).
156. J. G. Hellerman and D. A. Shafritz, *PNAS* **72**, 1021 (1975).
157. R. Kaempfer, R. Hollender, W. R. Abrams and R. Israeli, *PNAS* **75**, 209 (1978).
158. R. Kaempfer, H. Rosen and R. Israeli, *PNAS* **75**, 650 (1978).
159. A. Barrieux and M. G. Rosenfeld, *JBC* **252**, 392 (1977).
160. A. Barrieux and M. G. Rosenfeld, *JBC* **253**, 6311 (1978).
161. A. Barrieux and M. G. Rosenfeld, *JBC* **254**, 8087 (1979).
162. C. Baglioni, *BBA* **287**, 189 (1972).
163. S. L. Adams, B. Safer, W. F. Anderson and W. C. Merrick, *JBC* **250**, 9083 (1975).
164. D. T. Peterson, B. Safer and W. C. Merrick, *JBC* **254**, 7730 (1979).
165. A. Marcus, *JBC* **245**, 955 (1970).
166. R. Benne, J. Edman, R. R. Traut and J. W. B. Hershey, *PNAS* **75**, 108 (1978).
167. A. Efstratiadis, F. C. Kafatos and T. Maniatis, *Cell* **10**, 571 (1977).
168. J. A. Steitz and K. Jakes, *PNAS* **72**, 4734 (1975).
169. M. Kozak and A. J. Shatkin, *JBC* **252**, 6895 (1974).
170. S. Muthukrishnan, M. Morgan, A. K. Banerjee and A. J. Shatkin, *Bchem* **15**, 5761 (1976).
171. G. W. Both, Y. Furuchi, S. Muthukrishnan and A. J. Shatkin, *JMB* **104**, 637 (1976).
172. S. Muthukrishnan, B. Moss, J. A. Cooper and E. S. Maxwell, *JBC* **253**, 1710 (1978).
173. S. Legon, H. D. Robertson and W. Prensky, *JMB* **106**, 23 (1976).
174. M. Kozak, *Cell* **15**, 1109 (1978).
175. M. Kozak, *Nature* **280**, 82 (1979).
176. J. A. Steitz, in "Ribosome Structure, Function and Genetics" (G. Chambliss, G. R. Craven, J. Davies, K. Davis, L. Kahan, and M. Numura, eds.), p. 479. University Park Press, Baltimore, Maryland, 1980.
177. B. M. Rosenberg and M. Paterson, *Nature* **279**, 692 (1979).
178. B. M. Rosenberg and M. Paterson, *Nature* **279**, 694 (1979).
179. M. Kozak and A. J. Shatkin, *JBC* **253**, 6568 (1979).
180. W. Godchaux and K. C. Atwood, *JBC* **251**, 292 (1976).
181. R. Jagus, A. Koniesczny and B. Safer, manuscript in preparation.
182. A. Spirin, *EJB* **38**, 443 (1969).
183. G. Brawerman, *ARB* **43**, 621 (1974).
184. E. S. Gander, A. G. Stewart, C. M. Morel and K. Scherrer, *EJB* **38**, 443 (1973).
185. R. Williamson, *FEBS Lett.* **37**, 1 (1973).
186. O. Civelli, A. Vincent, J. F. Buri and K. Scherrer, *FEBS Lett.* **72**, 71 (1976).
187. A. Vincent, O. Civelli, J. F. Buri and K. Scherrer, *FEBS Lett.* **77**, 281 (1977).
188. S. K. Jain, M. G. Pluskal and S. Sarkar, *FEBS Lett.* **97**, 84 (1979).
189. S. K. Jain and S. Sarkar, *Bchem* **18**, 745 (1979).
190. S. K. Jain, R. K. Roy, M. G. Pluskal, D. E. Croall, C. Guha and S. Sarkar, *Mol. Biol. Rep.* **5**, 79 (1979).
191. R. K. Roy, A. S. Lau, H. N. Munro, B. S. Baliga and S. Sarkar, *PNAS* **76**, 1751 (1979).
192. A. Kumar and T. Pederson, *JMB* **96**, 353 (1975).
193. V. Lindberg and B. Sundquist, *JMB* **86**, 451 (1974).
194. V. M. Kish and T. Pederson, *JMB* **95**, 227 (1975).
195. T. J. Quinlan, P. B. Billings and T. E. Martin, *PNAS* **71**, 2632 (1974).
196. K. M. Rose, S. T. Jacob and A. Kumar, *Nature* **279**, 260 (1979).
197. A. Rourke and S. M. Heywood, *Bchem* **11**, 2061 (1972).

198. S. M. Heywood, D. S. Kennedy and A. J. Bester, PNAS **71**, 2428 (1974).
199. D. S. Kennedy and S. M. Heywood, FEBS Lett. **72**, 314 (1976).
200. F. Golina, S. S. Thach, C. H. Birge, B. Safer, W. C. Merrick and R. E. Thach, PNAS **73**, 3040 (1976).
201. D. Kabat and M. R. Chappell, JBC **252**, 2684 (1977).
202. S. M. Heywood and D. S. Kennedy, ABB **192**, 270 (1979).
203. W. R. Gette and S. M. Heywood, JBC **254**, 9879 (1979).
204. H. Trachsel and T. Staehelin, PNAS **75**, 204 (1978).
205. D. H. Levin, D. Kynes and G. Acs, JBC **248**, 6416 (1973).
206. L. M. Cashion and W. M. Stanley, PNAS **71**, 436 (1974).
207. H. Suzuki and I. H. Goldberg, PNAS **71**, 4529 (1974).
207a. F. Grummt, EJB **43**, 337 (1976).
208. W. C. Merrick, JBC **254**, 3708 (1979).
208a. H. Weissbach, in "Ribosomes: Structure, Function and Genetics" (G. Chambliss, G. R. Craven, J. Davies, K. Davis, L. Kahan, and M. Numura, eds.), p. 377. University Park Press, Baltimore, Maryland, 1980.
209. J. Kruh and H. Borsook, JBC **220**, 95 (1956).
209a. G. P. Bruns and I. M. London, BBRC **18**, 236 (1965).
210. A. I. Grayzel, P. Hörchner and I. M. London, PNAS **55**, 650 (1966).
211. H. S. Waxman and M. Rabinowitz, BBA **129**, 369 (1966).
212. H. S. Waxman, M. L. Freedman and M. Rabinovitz, BBA **145**, 353 (1967).
213. R. Levere and S. Granick, JBC **242**, 1903 (1967).
214. S. Sassa, S. Granick, C. Chang and A. Kappas, in "Erythropoiesis" (N. Nakao, ed.), p. 383. University Park Press, Baltimore, Maryland, 1974.
215. H. Morell, J. C. Savoie and I. M. London, JBC **233**, 923 (1958).
216. T. R. Rutherford and D. J. Weatherall, Cell **16**, 415 (1979).
217. S. Granick and S. I. Beale, Adv. Enzymol. **46**, 151 (1978).
218. C. Z. Yanoni and S. H. Robinson, Nature **258**, 330 (1975).
219. W. V. Zucker and H. M. Schulman, PNAS **59**, 582 (1968).
220. S. D. Adamson, E. Herbert and W. Godchaux, ABB **125**, 671 (1968).
221. M. Gross and M. Rabinovitz, BBA **299**, 472 (1973).
222. S. D. Adamson, E. Herbert and S. F. Kemp, JMB **42**, 247 (1969).
223. M. Gross and M. Rabinovitz, PNAS **69**, 1565 (1972).
224. M. Gross, BBA **366**, 319 (1974).
225. S. Legon, R. J. Jackson and T. Hunt, Nature NB **241**, 150 (1973).
226. M. J. Clemens, B. Safer, W. C. Merrick, W. F. Anderson and I. M. London, PNAS **72**, 1286 (1975).
227. M. J. Clemens, EJB **66**, 413 (1976).
228. K. Balkow, S. Mizuno and M. Rabinovitz, BBRC **54**, 315 (1973).
229. R. Jagus and B. Safer, unpublished observations.
230. W. Kemper, B. Safer and R. Jagus, FP **37**, 1942 (1978).
231. M. Gross, JBC **254**, 2378 (1979).
232. B. Safer and W. Kemper, unpublished observations.
233. M. Gross, JBC **254**, 2370 (1979).
234. C. R. Maxwell, C. S. Kamper and M. Rabinovitz, JMB **58**, 317 (1971).
235. C. R. Maxwell and M. Rabinovitz, BBRC **35**, 79 (1969).
236. G. A. Howard, S. D. Adamson and E. Herbert, BBA **213**, 237 (1970).
237. M. Gross and M. Rabinovitz, BBA **287**, 340 (1972).
238. M. Gross and M. Rabinovitz, BBA **299**, 472 (1973).
239. M. Gross, BBRC **57**, 611 (1974).

240. M. Gross, *BBA* **520**, 642 (1978).
241. S. Legon, A. Brayley, T. Hunt and R. J. Jackson, *BBRC* **56**, 745 (1974).
242. V. Ernst, D. H. Levin, R. S. Ranu and I. M. London, *PNAS* **73**, 1112 (1976).
243. K. Balkow, T. Hunt and R. J. Jackson, BBRC **67**, 366 (1975).
244. D. H. Levin, R. S. Ranu, V. Ernst, M. A. Fifer and I. M. London, *PNAS* **72**, 4849 (1975).
245. P. J. Farrell, K. Balkow, T. Hunt, R. J. Jackson and H. Trachsel, *Cell* **11**, 187 (1977).
246. D. H. Levin, R. Ranu, V. Ernst and I. M. London, *PNAS* **73**, 3112 (1976).
247. G. Kramer, M. Cimadevilla and B. Hardesty, *PNAS* **73**, 3078 (1976).
248. M. Gross and J. Mendelewski, *BBRC* **74**, 559 (1977).
249. P. J. Farrell, T. Hunt and R. J. Jackson, *EJB* **89**, 517 (1978).
250. B. Safer, M. Lloyd, R. Jagus and W. Kemper, *FP* **37**, 1938 (1978).
251. V. Ernst, D. H. Levin and I. M. London, *PNAS* **76**, 2118 (1979).
252. M. Gross and M. Rabinovitz, *BBRC* **50**, 832 (1973).
253. R. S. Ranu and I. M. London, *PNAS* **73**, 4349 (1976).
254. H. Trachsel, R. S. Ranu and I. M. London, *PNAS* **75**, 3654 (1978).
255. M. Gross and J. Mendelewski, *BBA* **520**, 650 (1978).
255a. T. Hunt, *in* "Miami Winter Symposium" (T. R. Russell, K. Brew, H. Faber and J. Schule, eds.), Vol. 16, p. 321. Academic Press, 1979.
256. A. Datta, C. deHaro, J. M. Sierra and S. Ochoa, *PNAS* **74**, 1463 (1977).
257. A. Datta, C. deHaro, J. M. Sierra and S. Ochoa, *PNAS* **74**, 3326 (1977).
258. A. Datta, C. deHaro and S. Ochoa, *PNAS* **75**, 1148 (1978).
259. S. Ochoa and C. deHaro, *ARB* **48**, 549 (1978).
260. N. Grankowski, G. Kramer and B. Hardesty, *JBC* **254**, 3145 (1979).
261. D. Levin, V. Ernst and I. M. London, *JBC* **254**, 7935 (1979).
262. B. Safer and R. Jagus, *PNAS* **76**, 1094 (1979).
263. B. Safer, R. Jagus, D. Crouch and W. Kemper, *in* "Protein Phosphorylation and Bioregulation" (G. Thomas, E. J. Podesta and J. Gordon, eds.), p. 142. EMBO Workshop, Karger, Basel, 1979.
264. D. Crouch and B. Safer, *JBC* **255**, 7918 (1980).
265. D. Crouch, A. Stewart and B. Safer, *FP* (in press).
266. P. Cohen, *Curr. Top. Cell. Regul.* **14**, 117 (1978).
267. E. Krebs and G. Beavo, *ARB* **49**, 924 (1979).
268. B. Safer, W. Kemper and R. Jagus, *JBC* **254**, 8091 (1979).
269. D. Levin, V. Ernst, A. Leroux, R. Petryshyn, R. Fagard and I. M. London, *in* "Protein Phosphorylation and Bioregulation" (G. Thomas, E. J. Podesta and J. Gordon, eds.), p. 128. EMBO Workshop, Karger, Basel, 1979.
270. B. Safer, R. Jagus, and D. Crouch, *JBC* **255**, 6913 (1980).
271. W. C. Merrick, D. T. Peterson, B. Safer, M. Lloyd and W. Kemper, *in* "Proceedings of the Eleventh FEBS Meeting," Vol. 43, p. 17. Elsevier, Amsterdam, 1977.
272. R. Jagus and B. Safer, *FP* **38**, 1244 (1979).
273. R. Jagus and B. Safer, *JBC* (in press, 1981).
274. V. Ernst, D. Levin and I. M. London, *JBC* **253**, 7163 (1978).
275. M. Gross, *BBRC* **67**, 1507 (1975).
276. M. Gross, *BBA* **447**, 445 (1976).
277. M. Gross, *ABB* **180**, 121 (1977).
278. R. O. Ralston, A. Das, M. Grace, H. Das and N. K. Gupta, *PNAS* **76**, 5490 (1979).
279. H. Amesz, H. Goumans, T. Haubrich-Moiree, H. O. Voorma and R. Benne, *EJB* **98**, 513 (1979).
280. R. O. Ralston, A. Das, M. Grace, H. Das and N. K. Gupta, *PNAS* **76**, 5076 (1979).

281. D. H. Metz, *Cell* **6**, 429 (1975).
282. R. M. Friedman, *J. Virol.* **2**, 1081 (1968).
283. D. H. Metz and M. Esteban, *Nature* **238**, 385 (1972).
284. W. K. Joklik and T. C. Merigan, *PNAS* **56**, 558 (1966).
285. D. H. Metz, M. Esteban and G. Danielescu, *J. Virol.* **27**, 197 (1975).
286. I. M. Kerr, *J. Virol.* **7**, 448 (1971).
287. R. M. Friedman, D. H. Metz, R. M. Esteban, D. R. Tovell, L. A. Ball and I. M. Kerr, *J. Virol.* **10**, 1184 (1972).
288. E. Falcoff, R. Falcoff, B. Lebleu and M. Revel, *J. Virol.* **12**, 421 (1973).
289. S. L. Gupta, M. L. Sopori and P. Lengyel, *BBRC* **54**, 777 (1963).
290. C. E. Samuel and W. K. Joklik, *Virology* **58**, 476 (1974).
291. I. M. Kerr, R. M. Friedman, R. E. Brown, L. A. Ball and J. C. Brown, *J. Virol.* **13**, 9 (1974).
292. J. Content, R. Lebleu, A. Zilberstein, H. Berissi and M. Revel, *FEBS Lett.* **41**, 125 (1974).
293. I. M. Kerr, R. E. Brown and L. A. Ball, *Nature* **250**, 57 (1974).
294. I. M. Kerr, R. E. Brown, M. J. Clemens and C. S. Gilbert, *EJB* **69**, 551 (1976).
295. W. R. Roberts, M. J. Clemens and I. M. Kerr, *PNAS* **73**, 3136 (1976).
296. A. Zilberstein, P. Federman, L. Shulman and M. Revel, *FEBS Lett.* **68**, 119 (1976).
297. W. K. Roberts, A. G. Hovanessian, R. E. Brown, M. J. Clemens and I. M. Kerr, *Nature* **264**, 477 (1976).
298. C. Baglioni, *Cell* **17**, 255 (1979).
299. A. G. Hovanessian and I. M. Kerr, *EJB* **23**, 515 (1979).
300. B. Lebleu, G. Sen, S. Shaila, B. Cabrer and P. Lengyel, *PNAS* **73**, 3107 (1976).
301. J. A. Cooper and P. J. Farrell, *BBRC* **77**, 124 (1977).
302. W. K. Roberts, A. Hovanessian, R. E. Brown, M. J. Clemens and I. M. Kerr, *Nature* **264**, 477 (1976).
303. C. E. Samuel, D. A. Farris and D. A. Eppstein, *Virology* **83**, 56 (1977).
304. D. Levin and I. M. London, *PNAS* **75**, 1112 (1978).
305. C. E. Samuel, *Virology* **93**, 281 (1979).
306. A. Kimchi, A. Zilberstein, A. Schmidt, L. Shulman and M. Revel, *JBC* **254**, 9846 (1979).
307. A. P. Jarvis, C. Shite, A. Ball, S. L. Gupta, L. Ratner, G. C. Sen and C. Colby, *Cell* **14**, 879 (1978).
308. J. R. Lenz and C. Baglioni, *JBC* **253**, 4219 (1978).
309. A. Zilberstein, A. Kimchi, A. Schmidt and M. Revel, *PNAS* **75**, 4734 (1978).
310. J. A. Lewis, E. Falcoff and R. Falcoff, *EJB* **86**, 497 (1978).
311. G. C. Sen, H. Taira and P. Lengyel, *JBC* **253**, 5915 (1978).
312. M. Revel, E. Gilboa, A. Kimchi, A. Schmidt, L. Shulman, E. Yakobson and A. Zilberstein, in "Proceedings of the Eleventh FEBS Meeting," Vol. 43, p. 47. Elsevier, Amsterdam, 1977.
313. J. Pichon, P. J. Farrell and P. Lengyel.
314. D. Levin, R. Petryshyn and I. M. London, *PNAS* **77**, 832 (1980).
315. R. Petryshyn, H. Trachsel and I. M. London, *PNAS* **76**, 1575 (1979).
316. C. E. Samuel, *PNAS* **76**, 600 (1979).
317. S. L. Gupta, *J. Virol.* **29**, 301, 1979.
318. A. G. Hovanessian and I. M. Kerr, *Nature* **265**, 537 (1977).
319. I. M. Kerr, R. E. Brown and A. G. Hovanessian, *Nature* **268**, 540 (1977).
320. I. M. Kerr and R. E. Brown, *PNAS* **75**, 256 (1978).
321. I. A. Ball and C. N. White, *PNAS* **75**, 1167 (1978).

322. C. Baglioni, M. A. Minks and P. A. Maroney, *Nature* **273**, 684 (1978).
323. L. Ratner, R. C. Wiegand, P. J. Farrell, G. C. Sen, B. Cabrer and P. Lengyel, *BBRC* **81**, 967 (1978).
324. A. G. Hovanessian and I. M. Kerr, *EJB* **84**, 149 (1978).
325. I. M. Kerr, B. R. G. Williams, R. R. Golgher, G. R. Stark and E. M. Martin, *in* "Protein Phosphorylation and Bioregulation" (G. Thomas, E. J. Podesta and J. Gordon, eds.), p. 169. EMBO Workshop, Karger, Basel, 1979.
326. M. A. Minks, S. Benvin, P. A. Maroney and C. Baglioni, *JBC* **6**, 767 (1979).
327. L. A. Ball, *Virology* **94**, 282 (1979).
328. J. P. Dougherty, H. Samanta, P. J. Farrell and P. Lengyel, *JBC* **255**, 3813 (1980).
329. M. J. Clemens and B. R. G. Williams, *Cell* **13**, 565 (1978).
330. C. M. Vaqueso and M. J. Clemens, *EJB* **98**, 245 (1979).
331. B. R. G. Williams, I. M. Kerr, C. S. Gilbert, C. N. White and L. A. Ball, *EJB* **92**, 455 (1978).
332. G. E. Brown, B. Lebleu, M. Kawakita, S. Shaila, G. C. Sen and P. Lengyel, *BBRC* **69**, 114 (1976).
332a. E. Slattery, N. Ghosh, H. Samanta and P. Lengyel, *PNAS* **76**, 4778 (1979).
333. B. R. G. Williams and I. M. Kerr, *Nature* **276**, 88 (1978).
334. E. Yakobson, C. Prives, J. R. Hartman, E. Vincour and M. Revel, *Cell* **12**, 73 (1977).
335. S. L. Gupta, W. D. Graziadei, H. Weidely, M. C. Sopori and P. Lengyel, *Virology* **57**, 49 (1974).
336. P. Lengyel, G. Sen, L. Ratner, P. Farrell, M. Dubois, B. Cabrer, H. Taira, E. Slattery, R. Weigand and R. Desioseiro, *Abstr. Int. Congr. Virol., 4th, 1978* p. 97 (1978).
337. T. W. Nilsen and C. Baglioni, *PNAS* **76**, 2600 (1979).
338. G. R. Stark, W. J. Power, R. T. Schimke, R. E. Brown and I. M. Kerr, *Nature* **278**, 471 (1970).
339. N. Shimizu and Y. Sokawa, *JBC* **254**, 12034 (1979).
340. M. Revel, communication at EMBO Workshop, Basel, 1979.
341. V. M. Pain and E. C. Henshaw, *EJB* **57**, 335 (1975).
342. V. M. Pain, *in* "Biochemistry of Cellular Regulation" (M. J. Clemens, ed.), Vol. I, p. 85. CRC Press, West Palm Beach, Florida, 1980.
343. A. Weber, E. R. Feman and C. Baglioni, *Bchem* **14**, 5315 (1975).
344. V. M. Pain, J. A. Lewis, P. Huuos, E. C. Henshaw and M. J. Clemens, *JBC* **255**, 1486 (1980).
345. M. H. Vaughan and B. S. Hansen, *JBC* **248**, 7087 (1973).
346. M. E. Smulson and M. Rabinovitz, *ABB* **124**, 306 (1968).
346a. M. Hori, S. M. Fisher and M. Rabinowitz, *Science* **155**, 83 (1967).
347. E. C. Henshaw, *in* "Modulation of Protein Function" (D. E. Atkinson and C. F. Fox, eds.), p. 407. Academic Press, New York, 1979.
348. M. J. Clemens, V. M. Pain, E. C. Henshaw and I. M. London, *BBRC* **72**, 768 (1976).
349. J. Delauney, R. S. Ranu, D. H. Levin, V. Ernst and I. M. London, *PNAS* **74**, 2264 (1977).
350. J. M. Sierra, C. Dettaro, A. Datta and S. Ochoa, *PNAS* **74**, 4356 (1977).
351. J. E. Kay, C. R. Benzie, P. Dicker and K. Lindahl-Kiessling, *FEBS Lett.* **91**, 40 (1978).

Structure, Replication, and Transcription of the SV40 Genome

GOKUL C. DAS[1,2] AND
SALIL K. NIYOGI

The University of Tennessee—Oak Ridge Graduate School of Biomedical Sciences and Biology Division, Oak Ridge National Laboratory,[3]
Oak Ridge, Tennessee

I. Introduction	187
II. Structure of the SV40 Genome	189
A. Biophysical Properties	189
B. Protein Composition	189
C. Repeat Length and Distribution of Nucleosomes	192
D. Heterogeneity	194
E. Artifacts of Chromatin Preparation	195
F. Models of Superstructure	199
III. Replication of the SV40 Genome	199
A. Replication of SV40 DNA	199
B. Gene Product(s) and Enzymes Involved in DNA Replication	203
C. Assembly of Histones on the SV40 Genome	207
IV. Transcription of the SV40 Genome	211
A. Transcription of SV40 Genome *in Vivo*	212
B. Transcriptional Control of Gene Expression	215
C. Posttranscriptional Control	223
V. The Minichromosome—A Model for the Structure and Function of Eukaryotic Chromatin	228
References	232
Note Added in Proof	240

I. Introduction

Over the last decade, several animal viruses have become quite useful as models for the understanding of molecular mechanisms of gene expression and regulation in higher organisms. Perhaps the most spectacular successes have been achieved with DNA-containing pa-

---

[1] Postdoctoral Investigator supported by subcontract No. 3322 from the Biology Division of Oak Ridge National Laboratory to the University of Tennessee.

[2] Present address: Laboratory of Biology of Viruses, NIAID, National Institutes of Health, Bethesda, Maryland 20014.

[3] Research sponsored by the Office of Health and Environmental Research, U. S. Department of Energy, under contract W-7405-eng-26 with the Union Carbide Corporation.

pova viruses (SV40, polyoma) and adenoviruses. Simian virus 40 (SV40) in particular has attracted wide attention because it can transform infected cells in living animals or cells in culture to tumorigenicity. The recent discovery that SV40 genome has a chromatin-like structure has further stimulated interest in this virus.

SV40 virions are about 500 Å in diameter with icosahedral symmetry of the outer shell (1, 2). It contains a duplex genome of about 5200 base-pairs, the sequence of which is now known (3, 4). The genome is complexed with the four histones H2A, H2B, H3, and H4 (5, 6) derived from host cells. The viral DNA codes for three capsid proteins, and for two or three tumor antigen proteins necessary for viral DNA replication and cellular transformation. Restriction endonucleases have played a crucial role in defining the genetic and functional organization of this genome (5, 6a). The cleavage sites provide coordinates for a molecular map of the DNA, and permit one to locate accurately particular physical features or loci for biochemical function. The single *E. coli* endonuclease (*Eco*RI) cleavage site serves as the reference point and is assigned map position 0/1.0. Apart from these, the SV40 genome is being used successfully as a vehicle for molecular cloning in eukaryotic cells (7).

Several excellent reviews on the molecular biology of this virus have appeared recently (8–11). The purpose of the present essay is to examine the current status of our knowledge of both the structure and the function of the SV40 genome, and to correlate these two aspects. Attempts have been made to understand replication and transcription mainly at the level of the DNA, but it now appears that the chromosome, rather than free DNA, is the actual template. Since polyoma virus belongs to the same group as does SV40 and has the same minichromosome-like structure, the results obtained with this virus will be considered here, if necessary, only to fill in the gaps of our knowledge of the SV40 systems.

A few different lines of evidence lead one to believe that the transcription of SV40 genes can provide an accurate model for transcription of genes in eukaryotic organisms. These include the facts that its transcription is initiated by RNA polymerase II (12), which is indistinguishable from the cellular enzyme, and that the posttranscriptional processing of SV40 mRNA is similar to that of cellular mRNAs, namely, capping, polyadenylylation, splicing, and internal methylation (13–15). Recent observations in this area are evaluated, not only for the understanding of the molecular biology of the virus, but also with respect to more general problems of transcription and replication in eukaryotic cells.

II. Structure of the SV40 Genome

A. Biophysical Properties

SV40 contains a circular double-stranded DNA of molecular weight 3.2×10^6 and having approximately 25 superhelical turns (reviewed in 5, 8). The whole genome, as a nucleoprotein complex, has been isolated from the nuclei of infected cells as well as from disrupted mature virions, in the form of a "beads-on-a-string" structure (16) akin to cellular chromatin (17). It is called the minichromosome because of its small size and chromatin-like appearance.

Depending on salt conditions and isolation procedures, minichromosomes can be obtained in different forms (Fig. 1) such as "beads-on-a-string" (16, 18, 19), 100 Å nucleofilaments (16), "superbeads" (20, 21), and as spherical compact particles (20, 22). At physiological ionic strength (~0.15 M), the minichromosomes isolated either from infected cells or from mature virions appear as spherical particles of diameter about 300 Å and with a sedimentation coefficient of about 75 S (22, 23). A higher ionic strength (about 0.6 M), which is sufficient to dissociate histone H1 or similar protein moieties, converts the minichromosome into a relaxed circle having about 20–24 nucleosomes and with a sedimentation coefficient of 35–55 S (20, 21). At intermediate ionic strengths, many other forms, such as 100 Å nucleofilaments (16) or superbeads, are possible (20, 21). Prolonged dialysis of compact minichromosomes against a buffer of a low ionic strength, for example, 1 mM TrisCl (pH 7.4), also yields a "beads-on-a-string" structure with different degrees of relaxation (19, 21); the reasons for this relaxation in the presence of H1 are not yet clear.

Superhelicity in deproteinized circular SV40 DNA is a consequence of the formation of nucleosomes (24). DNA containing about 25 negative superhelical turns can be obtained from either mature virions or from minichromosomes isolated from cell nuclei (21). This indicates that the DNA in each nucleosome is under a torsional constraint equivalent to about one negative superhelical turn, and that the higher-order folding of the minichromosome into the condensed state inside the virion does not require any nicking–closing event.

B. Protein Composition

SV40 virions contain three virus-coded structural proteins, VP1 (~70%), VP2 (~10%), and VP3 (~10%), and four host-specific histones, H2A, H2B, H3, and H4 (5, 6). The histones are present in

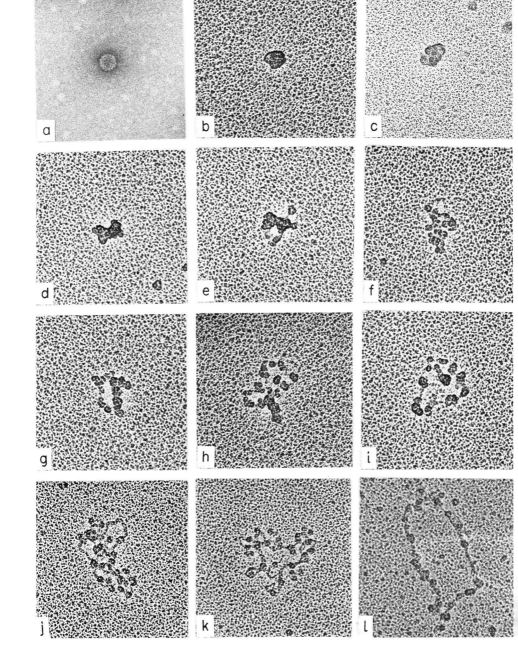

equimolar amounts in the nucleosome. The mass ratio of histone to DNA in the minichromosome, like cellular chromatin, is about 1.0 (18, 20). In early experiments, H1 was not found either in the mature virion or in the minichromosome (6, 25–27). Contrary to one report (28), there is now increasing evidence that both the intracellular mature minichromosome (20, 21, 29) and its replicating intermediates (30, 31) contain histone H1, but that mature virions do not (31–33). However, it is difficult to establish the presence of histone H1, as it migrates in dodecyl-sulfate electrophoresis at the same rate as does VP3 (34). The molar amount of histone H1 being about one-half that of the other histones, the amount of the material examined is also a critical factor for its detection in the minichromosome.

Besides cellular histones, minor quantities of capsid proteins are sometimes associated with the minichromosome in varying proportions when it is isolated either from the infected cells or from mature virions (21, 31). It is becoming clear that the capsid proteins play a dual role in the structure and function of the virions (35; G. Stark, personal communication) and discussed under transcription (Section IV,B). During the lytic cycle of SV40, the level of histone acetylation in the host chromatin does not rise significantly above that of the uninfected cells, but the intracellular viral minichromosome (75 S) contains histones more acetylated than the host histones (36). Histones H3 and H4 in SV40 and polyoma minichromosomes are more acetylated than histones H2A and H2B (32, 37, 38). The acetylation of nucleosomal histones increases progressively as the recently assembled minichromosome (75 S) gradually matures with the addition of capsid proteins, and ends up in the virion with highly acetylated histones (36, 39), but histone H1 is eliminated at this final stage of maturation (31, 39). It is possible that the observations (28) indicating the presence of histone H1 in mature virions were made on previrion structures (as discussed in detail in Section II,D) purified in a sucrose gradient and having the same sedimentation coefficient as that of mature virions (31, 39).

FIG. 1. Electron micrographs of mature virions (a) and of SV40 minichromosomes (b-l) isolated from mature virions as described in references 19 and 65. Highly compact and globular structures, beads-on-a-string, and selected intermediates with different degrees of relaxation under different conditions are shown to illustrate the transition of the compact globular structure to the beaded appearance of the SV40 minichromosome. Magnification in each case is 100,000. (G. C. Das, D. P. Allison, and S. K. Niyogi, unpublished observations)

C. Repeat Length and Distribution of Nucleosomes

The kinetics of action and the products of digestion of SV40 minichromosomes by microccocal nuclease (40, 41) are similar to those of cellular chromatin (41a). The repeat length thus determined is equal to about 200 base-pairs. Electron microscopy of the relaxed minichromosome indicates the presence of a variable stretch of relatively free DNA between nucleosomes (16, 18, 20, 21). When the length of the free DNA is measured, that embodied in the nucleosome can easily be calculated and is consistent with the repeat length obtained biochemically (18, 40).

The presence of varying lengths of nonnucleosomal DNA may be an artifact of chromatin preparation arising from loss or from the sliding of nucleosomes along the chromatin fiber. To confirm this idea, both the SV40 and cellular chromatin were digested inside nuclei isolated from SV40-infected cells (42). The average repeat length for the DNA oligomers from the SV40 minichromosome is 187 ± 11 base-pairs, which differs only slightly from that observed for the host CV-1 chromatin (182 ± 6 base-pairs). If the minichromosome were composed of uniformly spaced nucleosomes with an average repeat length of 187 base-pairs of DNA, 28 nucleosomes could be accommodated on the entire genome as against 21 ± 2 observed in the isolated minichromosome. A simple probability model (42a) shows that the SV40 genome cannot contain more than 22.3 ± 0.8 nucleosomes if nucleosomes of the same size are distributed randomly along the genome in a manner such that the internucleosomal distance never exceeds the length of the DNA associated with a single nucleosome. This calculation leaves about 19% of the genome as nonnucleosomal DNA.

The validity of the assumption of random distribution of nucleosomes can be questioned. One method of examining this is to probe the availability of known restriction endonuclease cleavage sites in the chromatin. If all sites are equally available, the arrangement is random; if some are protected, the arrangement is nonrandom. The first few reports on this issue tended to favor a random location of nucleosomes (43, 44), but there are other reports indicating the opposite (19, 45, 46). A recent report, however, supports a nearly random distribution of nucleosomes with respect to DNA sequences throughout the genome and a nearly random phasing between the chromosome structures of sibling molecules (47).

Another kind of specificity of nucleosomal arrangement has been reported by several laboratories. A nuclease-sensitive and exposed region of about 75–400 base-pairs is located in the "late" side of the origin of replication as probed by the single-cut restriction endonuclease BglI, by DNase I or by endogenous endonucleases (48–51). This region does not code for any known protein, but contains the heterogeneous 5' ends of the late SV40 mRNAs. Viral mutants containing duplicated DNA segments, including the origin of viral DNA replication and contiguous regions, have been used to define *cis* genetic elements responsible for the endonuclease-sensitive structure adjacent to the origin of replication (D. J. Wigmore, R. W. Eaton, and W. A. Scott, personal communication). More than one *cis* genetic element appears to be involved in the organization of this structure.

Under the electron microscope, a gap devoid of nucleosomes and located in the late side of the origin of replication is observed in about 20% of the relaxed viral chromosomes isolated from infected cells (52, 52a). Since only 20% of the molecules contain this gap, it is not clear whether it reflects a structural heterogeneity related to a specific function of the minichromosome. To elucidate this point, it would be interesting to study the presence of this gap in the minichromosomes isolated from mature virions.

The primary cleavage sites of microccocal nuclease at the supranucleosomal level of the minichromosome is nonrandomly and unevenly distributed in several clusters (53, 54), the shortest distance between two neighboring abundant sites is about 140–150 base-pairs (54). However, about half of all cuts are concentrated at the origin of replication and in the nearby late portion of the SV40 genome (53). But in one form of the compact minichromosome, the origin of replication is not exposed to the single-cut restriction endonuclease BglI (19). This apparent discrepancy may arise from the specific arrangement of the nucleosome-free region in the compact minichromosome, or perhaps other protein(s), such as T antigen, cover this region under the conditions employed. This is consistent with the presence of a dodecyl-sulfate and salt-stable protein associated with the SV40 DNA near the origin of replication, as observed under the electron microscope (55, 56). However, other factors may be responsible for the variation of the endonuclease cleavage results: (a) different material used by different workers; (b) cleavage by the endogenous endonuclease associated with chromatin (51); and (c) the SV40 minichromosome in a population may be heterogeneous (57–61) (see Section II, D).

D. Heterogeneity

During the period between the uncoating of the virus in infected cells and the formation of progeny virus particles, the viral minichromosome passes through distinct stages including early transcription, replication, late transcription, and association of the capsid proteins, leading finally to the maturation of the virus particles. These events, spanning more than one generation of the viral genome, should lead to structural heterogeneity of the minichromosome in the nuclei of infected cells and hence in a population at some definite time after infection. In the past, this possibility was either overlooked or ignored when the minichromosome was isolated from infected cells under different experimental conditions (16, 18–22).

With improvement in the isolation and purification procedures, it is now possible to extract several different forms of SV40 nucleoprotein complexes distinguishable by their labeling kinetics and apparent sedimentation constants (57–59). A relatively heterogeneous replicative intermediate of about 75–200 S is converted to a 75 S complex shortly after completion of replication. At least a portion of the 75 S chromatin is rapidly assembled into 200 S nucleoprotein complexes and subsequently into a 240 S previrion form that resembles mature virions (57–61a). The previrion has the same sedimentation coefficient and buoyant density as the mature one, but differs in stability, size, and infectivity (60, 61). At the final stage of maturation, histone H1 is eliminated from the previrion and probably replaced by VP3 (31, 39). The previrion dissociates at high ionic strength into empty capsid, nucleoprotein complex, and soluble proteins (61), and it is unlikely to be the degradation product of the mature virion. It appears that the empty capsids, generally found in CsCl gradients, are the degradation product of the previrion structure and not the virus precursor. Therefore, in the assembly process, viral chromatin is not introduced into preassembled empty capsids, as proposed earlier (63); rather, capsid proteins are gradually added to a viral minichromosome to form a salt-labile encapsidated structure that matures to a stable virus particle (39, 58, 61).

Divalent cations, such as Ca^{2+}, may have a role in stabilizing the salt-labile structure, since chelation of Ca^{2+} by EGTA disrupts the virus (64). In cells infected with tsB-11 mutant (which is defective in the production of VP1, a major structural protein), the 75 S nucleoprotein complex accumulates gradually, and no previrion or mature virion is assembled at the restrictive temperature (59). It is not known whether, in the absence of VP1, the other two structural proteins can bind to the nucleoprotein complex. Thus, cells infected with tsB-11 may be a good source for a homogeneous nucleoprotein complex and offer an ideal opportunity to examine whether VP1 has any role in the late gene expression of SV40.

E. Artifacts of Chromatin Preparation

The procedures for extracting and purifying minichromosomes are very important for obtaining a native, homogeneous population. The extraction methods employed in the past seem to have yielded complexes that arose, in part, from the disruption of mature virions already in the nucleus (57–62). Electron microscopy of cellular extracts in the presence of Triton X-100 frequently show partially disrupted virions with different amounts of minichromosome emerging from them (G. C. Das, unpublished results). Of all the factors affecting the stability of SV40 virions during the extraction of minichromosomes from infected cells, the most important ones seem to be the high ionic strength and chelating agents (62, 64). The compact form of the minichromosome can be readily converted to the relaxed 55 S complex by simple variations of the isolation procedures, such as longer extraction time, high ionic strength, and incubation at 37°C. Thus, in many instances the samples under study have been heterogeneous and/or devoid of superstructure; as a result, conclusions drawn on the basis of biochemical or enzymological studies need reevaluation.

A significant amount of protein is lost from minichromosomes isolated from mature virions by exposure to high pH (e.g., pH 10.5) (41, 43). With modification of the original method (45), it is possible to get a highly compact minichromosome of about 110 S from mature virions (19, 65). Another type of nucleoprotein complex is obtained with chelating agents (35, 64), which dissociates the virus into a core nucleoprotein complex and protein capsomers. Both nucleoprotein complexes have the same S value (37, 65), and the same viral capsid proteins (but of varying proportions) are associated with them (61; G. C. Das and S. K. Niyogi, unpublished results). In the latter case, the infectious virus particle could be reconstituted from the dissociated components (66). An outline of several isolation procedures adopted by some authors and the biophysical and biochemical properties of the isolated minichromosomes are presented in Table I. As seen in Table I, in most of the earlier studies, the minichromosome obtained either from infected cells or from the mature virions, was a 55 S nucleoprotein complex (16); in some cases, the sedimentation coefficient was as low as 30 S (43). Thus it appears

TABLE I

ISOLATION AND PURIFICATION OF SV40 MINICHROMOSOMES, AND THEIR BIOPHYSICAL AND BIOCHEMICAL PROPERTIES

Buffers used for isolation	Properties of the isolated minichromosome	References
A. From infected cells; the procedure is divided into three basic steps: I, isolation of nuclei; II, extraction of minichromosomes; III, purification in sucrose gradients		
1. I. 10 mM TrisCl (pH 7.4), 10 mM NaHSO$_3$, and 0.15 M NaCl II. Same buffer but also containing 0.25% Triton X-100, and incubated for 5 minutes at 37°C III. In buffer as in I	50–60 S; in 0.15 M NaCl, nucleoprotein complex appears as a nucleofilament of 105 ± 10 Å in diameter. In 0.015 M NaCl, beads-on-a-string structure containing 21 ± 1 nucleosomes of diameter 110 Å. DNA in each nucleosome (repeat length) is about 170 base-pairs. Length of the spacer is about 40 base-pairs. Packing ratio of DNA ≅ 7.	(16)
2. I. 10 mM TrisCl (pH 7.9), 10 mM EDTA II. 10 mM TrisCl (pH 7.9), 1 mM EDTA, 0.25% Triton X-100, 0.2 M NaCl III. 10 mM TrisCl (pH 7.9), 1 mM EDTA, 0.2 M NaCl	55 S; relaxed circular molecule containing about 20 nucleosomes of diameter 125 Å; repeat length ≅ 175–205 base-pairs	(18)
3. I. 10 mM TrisCl (pH 7.9), 5 mM EDTA II. 10 mM TrisCl (pH 7.9), 200 mM NaCl, 1 mM NaHSO$_3$, 0.25% Triton X-100 III. 10 mM TrisCl (pH 7.9), 0.2 mM EDTA, 0.2 M NaCl, 1 mM S$_2$threitol, and 1 mM NaHSO$_3$	55 S; typical beads-on-a-string structure containing 20 ± 2 nucleosomes; DNA repeat length ≅ 190–200 base-pairs. Nucleosomal core contains about 135 base-pairs of DNA. Condensation of the relaxed structure occurs in presence of histone H1.	(40)
4. I. 10 mM TrisCl (pH 7.8), 10 mM EDTA, 0.15 M NaCl, 0.25% Triton X-100, 10 minutes at 4°C II. Same as in I, kept at 4°C for 2–3 hours with gentle stirring with a glass rod III. 0.15 M NaCl, 1 mM NaEDTA, 1 mM TrisCl (pH 7.8)	55–60 S; "beads-on-a-string" structure containing histone H1. When molarity of NaCl is reduced to 0.13 M and the pHs of TrisCl are changed to 6.8 (I), 8.0 (II), and 7.5 (III), minichromosome is 75 S, compact, 300 Å diameter, and resistant to micrococcal nuclease	(22,29)

5.		Same as in (4) with the following modifications. pH used: I, 6.8; II, 8.0; III, 7.5. NaCl, 0.12 M throughout	Two peaks were obtained in the sucrose gradient. Peak I: Virions, 450 Å diameter, $\rho = 1.345$ g/ml Peak II: Compact minichromosome, 70 S, 300 Å diameter	(28)
6.	I.	0.2 mM potassium phosphate (pH 7.5), 1 M sucrose, 2 mM EDTA, NP-40	Materials in both peaks contain histone H1. 70 S; superbeads (0.15 M NaCl); H1 and varying amounts of VP1, VP2, VP3 are present. Relaxed structure containing about 24 nucleosomes (0.6 M NaCl).	(21)
	II.	20 mM Hepes (pH 7.8), 5 mM KCl, 0.5 mM MgCl$_2$, and 0.5 mM S$_2$threitol		
	III.	20 mM TrisCl (pH 7.8), 50 mM (NH$_4$)$_2$SO$_4$ containing varying molarity of NaCl from 50 to 600 mM		
7.	I.	10 mM Pipes (pH 6.8), 100 mM KCl	Heterogeneity of the minichromosomes: (i) 75–200 S (replicative intermediates), (ii) 75 S (mature minichromosome), (iii) 200 S (previrion), (iv) 250 S (mature virion)	(59)
	II.	Same as in I containing 1% Triton X-100, 0.5% Brij-58		
	III.	10 mM Pipes (pH 6.8), 100 mM KCl, 1 mM EDTA		
8.	I.	25 mM TrisCl (pH 7.4), 0.136 M NaCl, 7 mM KCl, 0.7 mM Na$_2$HPO$_4$, 0.5% NP-40	Heterogeneity of the minichromosomes: (i) 70 S, (ii) 180 S, (iii) 210 S. (i) and (ii) contain histone H1, but (iii) does not. Last two forms contain capsid proteins.	(31,58)
	II.	Same as in I but in absence of NP-40		
	III.	50 mM TrisCl (pH 7.9)		
9.	I.	10 mM Hepes (pH 7.8), or Pipes (pH 7.5), 5 mM KCl, 1 mM S$_2$threitol, 0.5 mM MgCl$_2$, 1 mM αTosF	Heterogeneity of the minichromosomes: (i) 95 S (replicating molecules), (ii) 75 S (mature minichromosomes), (iii) 200 S, (iv) 240 S previrion, (v) 240 S mature virions. All forms except (v) contain histone H1.	(57,61)
	II.	As in I		
	III.	As in II		
10.	I.	50 mM TrisCl (pH 7.9), 1 mM MgCl$_2$, 5 mM 2-mercaptoethanol	Heterogeneity of the minichromosome: (i) 95 S (replicating molecules), (ii) 75 S (mature minichromosome), (iii) 250 S (mature virions)	(52a, 61a)
	II.	Same as in I, incubated at 37° for 30 minutes		
	III.	Same buffer as in I but containing 100 mM (NH$_4$)$_2$SO$_4$		

(Continued)

TABLE I (Continued)

Buffers used for isolation	Properties of the isolated minichromosome	References
B. From mature virions; the procedure is divided into two basic steps: I, disruption of the virion; II, purification of the minichromosome in sucrose gradients		
1. I. Isotonic Tris-ethanolamine (pH 10.5), (150 mM NaCl, 1 mM EDTA, 20 mM TrisCl, pH 7.2; pH was adjusted to 10.5 with undiluted ethanolamine). II. Neutral sucrose gradient	30 S; VP3 is associated with the nucleoprotein complex	(43)
2. I. 150 mM TrisCl, pH 9.3, 20 mM S_2threitol (freshly prepared), 10 mM EDTA incubated at 4°C for 1.4 hour II. (a) 50 mM TrisCl (pH 8.1); or (b) 10 mM TrisCl (pH 7.5), 60 mM KCl, 1 mM EDTA, or (c) 10 mM TrisCl (pH 7.5), 100 mM NaCl, 1 mM EDTA	50–60 S; digestion pattern with micrococcal nuclease shows bands for monomer and higher multimers together with a few submonomer bands	(45)
3. I. 1 mM EGTA, 3 mM S_2threitol, 0.15 M NaCl and 50 mM TrisCl (pH 8.5), 30 minutes at 32°C II. 5 mM EGTA, 3 mM S_2threitol, 150 mM NaCl, 10 mM TrisCl (pH 7.8), and 0.05% Triton X-100	110–115 S; $\rho = 1.40$ g/ml, contains about 34% of the total viral protein	(35)
4. I. 150 mM TrisCl (pH 9.3), 20 mM S_2threitol, 10 mM EDTA, 1 mM αTosF, 100 mM NaCl, 3 hours at 4°C followed by 2 minutes at 37°C II. 10 mM TrisCl (pH 7.4), 1 mM EDTA, 1 mM αTosF and containing either 0.15 M NaCl or 0.6 M NaCl	At 0.15 M NaCl, 100–110 S, compact structures, VP1, VP2, VP3 are associated with the minichromosome; at 0.6 M NaCl, 50–60 S Predominantly VP1 is associated with the minichromosome.	(19, 65)

that artifacts in chromatin preparation are introduced mainly by the heterogeneity of the population, resulting from the loss of superstructure, and by differences in protein composition of the minichromosome leading to significant variation in the biochemical or biophysical data.

F. Models of Superstructure

It is evident that, of all the different states in which the minichromosome may exist, the compact one is very close to the native encapsidated state. The circular dichroism signal of either mature virions or compact minichromosomes isolated from virions is about the same as that of nucleosome cores obtained by extensive digestion of cellular chromatin with micrococcal nuclease (65). This indicates that the linker and the nonnucleosomal DNA of the native minichromosome is highly condensed, with a circular dichroic ellipticity similar to that of the core DNA (65). It is still not well elucidated how the whole genome, containing about 24 nucleosomes of 125 Å diameter each, is organized in the virus capsid of about 500 Å in diameter. One possibility is that the superstructure is solenoidal with either a constant or a variable number of nucleosomes per turn, as proposed for cellular chromatin (67). It is then necessary to have a returning DNA strand because of the circular genome (22). The stretch of DNA of about 400 base-pairs around the origin of replication that contains no nucleosomes may act as the returning DNA (49, 52, 52a). Another possibility is that the nucleosomes form a dodecahedron (68), where it is not necessary to postulate a return DNA. The diameter of the sphere surrounding the dodecahedron is about 300 Å, in good agreement with the diameter of the compact minichromosome or with the internal volume of the SV40 capsid. However, it is important in constructing a model to take into account that higher-order structural transition is at least equally important to the alteration in the nucleosomal structure in the actively transcribing chromatin.

III. Replication of the SV40 Genome

A. Replication of SV40 DNA

1. INITIATION AND TERMINATION OF REPLICATION

DNA replication in SV40 begins at a unique site near map (Fig. 2) position 0.67, the latter being measured in the clockwise direction from the single *Eco*RI cleavage site (69, 70). Replication proceeds in both di-

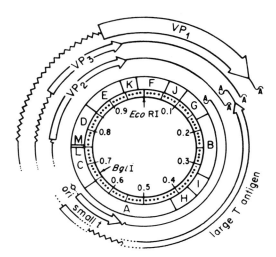

FIG. 2. Standard physical map of SV40 DNA and localization of main biological functions. The single cleavage site of the restriction enzyme EcoRI is used as a reference point for the physical map (inner circle). The next circle shows the position of the HindII and HindIII restriction fragments A to M. The origin of DNA replication (ori) is at or close to 0.663; this site corresponds to a large palindromic sequence, part of which represents the recognition sequence of BglI. The center of this palindrome is taken as the zero point for presenting the early (5' → 3' in counterclockwise orientation) and the late (5' → 3' in clockwise orientation) regions. The five virus-coded proteins are indicated by blocked arrows; untranslated parts of the mRNA are shown as solid lines; dots indicate uncertainty as to the exact position of the 5' end. Zigzag lines are used for the segments spliced out. A wavy line with A illustrates the 3'-terminal poly(A) tail. Reproduced from Fiers et al. (4, Fig. 1) with permission.

rection at about equal rates, and terminates at about 180° from the origin in the Hind G fragment of the physical map. These conclusions were drawn independently on the basis of two classical experiments: (a) the direct visualization under the electron microscope of replicating molecules containing a bubble and linearized by the cleavage of EcoRI under the electron microscope (69); and (b) the distribution of labels in the fragments produced by restriction endonucleases from pulse-labeled DNA (70).

The exact location of the origin of replication is not yet known, but the boundaries of the replication origin (ori) have recently been narrowed down to 85 nucleotides within 0.652–0.668 map unit (71, 72). The segment containing the origin of replication is characterized by a palindromic sequence of 27 base-pairs, rich in G · C, whose twofold axis of symmetry may be a characteristic of bidirectional replication (71, 73). The origin of replication appears to be a specific nucleotide signal ("ori-signal"). From a study of DNA synthesis in mutants ("ori-mutants") constructed by single base-pair substitutions in the palindromic sequence (74, 75), it appears that nucleotide sequences at positions 5154, 5155, 5158, 5159, and 5162 appear to be a part of the ori-signal, indicating that no unique base-pair is essential for viral DNA replication, but that the rate of DNA replication depends on the precise sequence at the "ori" site (75). Exact sequence symmetry is not an essential requirement, even for the bidirectional replication. A virus-coded early protein, T antigen, must interact with the origin of replication to initiate a replication cycle, but once the cycle is started, the T antigen is not required for continued synthesis (76–78). Still, very little is known about the exact mechanism of the initiation of replication.

Unlike the origin of replication, there is no unique sequence for termination, which occurs when the replication forks meet 180° from the initiation site (79). Some defective DNAs of both SV40 and polyoma appear to lack the portion of the viral genome in which termination normally occurs (80–82), and in some defective molecules, the replication fork passes through the region of *Hind*G containing the normal termination site (83). In ev-1114, a variant of SV40 in which a segment of the DNA containing the origin is duplicated, replication can begin at either origin, but always terminates approximately 180° around the circle from where it began (84).

When the two replicating forks meet at the termination site, two interlocked circles of the viral DNA are formed. These molecules segregate and give rise to two "component-II" DNA molecules (85). Each component-II DNA molecule consists of a strand of closed circular parental DNA and a daughter strand that contains a specific gap at position 0.17 (85, 86) of about 22–73 nucleotides (87). This is probably repaired with the help of a DNA polymerase and a ligase (86, 87). Although the exact mechanism of segregation is not known, it appears that the process comprises a scission in one template strand and an unwinding of this strand relative to the other, followed by gap filling and ligation. Failure of this process very often gives rise to a catenated dimer (88).

The overall replication time of the SV40 genome for a single round of synthesis (10–15 minutes) is about 10–20 times slower than that expected from mammalian chromosome replication. Late-replicating intermediates accumulate in at least two specific stages late during replication both *in vivo* and *in vitro* (89–91) indicating that the overall process of replication is discontinuous. Procedures that extract early replicating molecules fail to isolate the late replicating ones, which indicates that the latter are probably anchored within the nuclear matrix (89), as recently suggested for eukaryotic chromatin (92).

2. Mechanism of Chain Elongation

In both prokaryotes and eukaryotes, discontinuous synthesis of one of the strands seems obligatory in the absence of a DNA polymerase capable of elongating polynucleotide chains in a $3' \rightarrow 5'$ direction (93, 94). SV40 DNA chain elongation also occurs in a discontinuous fashion (95). Upon denaturation, the pulse-labeled replicative intermediates yield two classes of growing DNA chains differing in size, namely, a heterogeneous class ranging from 6 S to 16 S (corresponding to one complete strand), and a rather discrete class of 4–5 S fragments (about 150 nucleotides long). Since the latter could be "chased" into mature viral DNA, they were thought to be intermediates in DNA chain elongation (95). It is controversial whether the 4 S fragments are synthesized from one or both strands of the double helix at each replication fork.

Two kinds of experimental evidence support discontinuous synthesis of both strands: (a) incorporation of more than 50% of the radioactivity into Okazaki pieces after a short labeling period (95, 96); (b) extensive self-annealing (70–90%) exhibited by isolated fragments (97–99). However, kinetic analysis of the appearance of radioactivity in short and long chains shows that the amount of label in short chains never exceeds that in long ones even at very short pulse times, and that, at pulse times shorter than the lifetime of 4 S fragments, the amounts of radioactivity in short and long chains are approximately equal (100, 101). This observation strongly argues that the DNA synthesis is semidiscontinuous, i.e., the synthesis of 4 S DNA fragments occurs from only one strand of DNA at each replication fork, and the second strand is extended in a continuous manner. In some cases, a small fraction of the Okazaki pieces always hybridizes with the potentially continuous strand (102), which may arise from a number of reasons, including a low frequency of unidirectional replication from the normal replication origin or from an abnormal one (103, 104), contamination of Okazaki pieces by short continuously synthesized strands, and repair of newly synthesized DNA (105, 106). However, it is now well accepted that DNA synthesis at each of the replication forks proceeds in a predominantly continuous manner on the forward arm, where the direction of chain elongation ($5' \rightarrow 3'$) and that of the fork movement are the same, but discontinuously on the retrograde arm (102, 107, 108).

Specific molecular events and enzymological details in the initiation, elongation, and maturation of the Okazaki pieces still remain unclear. At least 50% of the Okazaki pieces contain uniquely sized oligoribonucleotides, about 10 residues in length, covalently attached to their 5' ends. This RNA acts as a primer and is not synthesized at unique sites on the DNA template (99, 109–112). The primer RNA is excised at the same rate as the Okazaki pieces are joined to the growing daughter strands, and the gap thus created between the 4 S fragments and the longer growing chains is filled, probably by a DNA polymerase and a ligase (113). It appears that DNA polymerase α is involved in both gap filling and continuous growth of the chain (114, 115).

However, the Okazaki pieces in replicating E. coli are about 10 S (116), much larger than that found in SV40. In the latter system, the replication fork must proceed over nucleosomal histones, which may provide a clue for both the slower rate and smaller size of the Okazaki pieces (112). It is not clear whether there is any relation between the initiation signals of the 4 S fragments and the positioning of the nucleosomes. However, a recent study (47) indicates that initiation of Okazaki fragment synthesis does not occur at randomly chosen DNA sequences, but at 100–300 preferred sites throughout the genome. Digestion of replicating SV40 chromatin with S1 nuclease releases about 50–60% of the Okazaki fragments as 4–6 S duplex DNA, and 90% of these fragments are not contained in nucleosomes (117).

B. Gene Product(s) and Enzymes Involved in DNA Replication

1. T ANTIGEN

The "early" region of the SV40 genome codes for two proteins, the "large T" and "small t" antigens (118–120), with apparent molecular weights of 90,000–100,000 and 15,000–20,000, respectively. Small t shares methionine-containing tryptic peptides with large T, the two polypeptides being coded by a common 5' nucleotide sequence (119). Genetic evidence indicates that large T is directly involved in the initiation of viral DNA replication (76–78) while small t is probably needed as a promoter for efficient virus-induced cell transformation (121–123). The large-T antigen, besides its unique role in DNA replication, is involved in the stimulation of cellular DNA and RNA synthesis, neoplastic transformation, negative control of early transcription, and possibly "turning on" late transcription (76, 124–132). The exact mechanisms of these functional roles of large T are still unclear.

One possibility is that large T, which is highly phosphorylated

(*133*), acts as a specific DNA binding protein (*134–137*) and may initiate viral DNA replication by interacting directly with the regulatory sequences at the origin of replication. The strength of binding of the T antigen depends on the sequence of the DNA (*138*). Large T, purified from a defective hybrid of adeno and SV40 viruses, consists of a polypeptide chain that is largely encoded by sequences of SV40 gene A but that contains a small stretch (10,000–12,000) of an adenovirus-coded unknown polypeptide at its amino terminus in place of small t of the SV40 (*139*). This hybrid protein binds in a sequential manner to three specific sites contained within a region of about 120 nucleotides that encompass the origin of DNA replication and the start of early and late transcription (*137*). The second site mutations that suppress the replication defect of the "ori-mutants" have been localized by *in vitro* recombination or by marker rescue experiments to the gene for the T antigen, indicating that a direct T antigen-origin interaction plays an important role in the regulation of viral DNA replication (*140*).

Among other possibilities, large T may catalyze certain biochemical reactions required to initiate DNA synthesis. ATPase and protein kinase activities have been reported to be associated with large T (*139*). Further studies indicate that large T is an ATPase, not a protein kinase, the latter activity arising from a strongly associated but separate host protein (*141a–143*). The ATPase activity of large T is stimulated preferentially by single-stranded polymers, especially poly(dT) (*143*). It is tempting to correlate this property with the finding that the origin of replication in the supercompact form of the SV40 minichromosome is not readily available to the restriction endonuclease *Bgl*I, which recognizes it (*19*). It is possible that large T, upon binding to the origin of replication, utilizes its ATPase activity to unwind portions of duplex DNA locally, thereby allowing polymerases to initiate replication.

Large T is associated with both mature and replicating SV40 minichromosomes when isolated from infected cells (*77, 144*). That at least a part of the association is tight and specific was demonstrated by removing the histones from immunoprecipitated chromatin with dextran sulfate and heparin and digesting the DNA with *Hin*fI and *Hpa*II. The site of association was located within the fragment of 0.641 to 0.725 on the SV40 physical map that includes the origin of replication (*144*). Large T is present in the minichromosomes isolated from cells infected with the temperature-sensitive mutant *ts*A 58 and grown only at the permissive temperature, thereby emphasizing its role in the initiation of SV40 replication at the level of minichromosome (*77*). A study of the binding *in vitro* of large T with the minichromosome shows that about two to three times more large T is needed to bind to chromatin compared to naked DNA (46).

2. DNA POLYMERASES

Since SV40 does not code for its own DNA polymerase, it is believed that host cell enzymes are involved in viral replication. Nuclear extracts of host cells contain at least three different types of DNA polymerases, namely, α, β, and γ, but their specific roles in the replication of the SV40 genome is not well understood (145). Three different approaches have implicated DNA polymerase α in the replication of SV40 DNA. First, a severalfold increase in DNA-polymerase-α activity consistently appeared at the onset of DNA synthesis, while DNA-polymerase-β activity remained relatively constant (145–147) and results on DNA polymerase γ were variable (148). Second, although both DNA polymerases α and γ are associated with SV40 nucleoprotein complexes, polymerase α is specifically associated with the replicating rather than the nonreplicating minichromosome (115, 149–157). The presence of a minor activity of polymerase γ with the SV40 nucleoprotein complex (115, 150, 151) may be due in part to the contamination, since the effect of inhibitors of DNA synthesis on this polymerase activity rules out its involvement in SV40 replication (152). DNA polymerase β is not detected in SV40 nucleoprotein complexes, although its presence was demonstrated in the nuclear extract (150–152).

The last line of evidence for the involvement of polymerase α comes from studies on the effect of several inhibitors of DNA synthesis. Earlier studies with hydroxyurea and fluorodeoxyuridine, which diminish dNTP pools (153), and with arabinosyl NTPs, which preferentially inhibit DNA polymerase α (153, 154), showed an accumulation of Okazaki fragments (113, 155, 155a). This result was interpreted as the involvement of two different DNA polymerases in DNA replication—one in the synthesis of Okazaki fragments, and the other for gap filling (113, 155). But other investigations did not find any accumulation of Okazaki fragments in the presence of araC and araCTP (156). On the other hand, DNA polymerase α is relatively resistant to d_2TTP, compared to the β and γ polymerases, which are quite sensitive (157). Since d_2TTP or its precursor d_2T do not have any effect either on the multiplication of SV40 virions or on the synthesis and joining of Okazaki fragments in replicating SV40 DNA, it appears that DNA polymerase α is exclusively responsible for viral DNA replication (152, 157).

The roles of DNA polymerases (or other proteins) in DNA replication can be determined by studying the effects on *in vitro* DNA replication of selective depletion of the polymerase, and of the addition of purified polymerase to selectively depleted preparations. When DNA polymerase α is selectively removed from replicating SV40 chromosomes by extraction at pH 10 or by agarose gel filtration, the capability of the chromosomes to synthesize DNA *in vitro* is not reduced in proportion to the loss of polymerase α (*150*). The loss of the ability to synthesize *in vitro* was greater for putative continuous strand synthesis than for discontinuous strand synthesis. This may indicate that polymerase α is responsible for continuous strand synthesis or that some specific factors involved in the synthesis are lost during agarose gel filtration. Polymerase α purified from monkey cells cannot complete Okazaki fragments on purified replicating SV40 DNA (*47*). This indicates the loss of a cofactor(s) during purification.

3. OTHER ENZYMES AND STIMULATORY FACTORS

DNA replication in many prokaryotic and eukaryotic systems requires RNA priming (*99, 109–112, 158*). Although several RNA polymerases are present in mammalian cells, none has yet been proved to be involved in DNA replication. Replication of polyoma DNA is insensitive to either low or high concentrations of α-amanitin (*159*), thereby excluding the involvement of RNA polymerases II and III in this process (*160*). The possible involvement of nuclear RNA polymerase I is not proved yet, but, instead, a mammalian primase that resembles the dna G product of *E. coli* catalyzes the initiation of Okazaki pieces in polyoma DNA (*161*).

During chain elongation, the primer RNA must be replaced by deoxyribonucleotides, which involves the digestion of RNA followed by gap filling. While the latter reaction is catalyzed by one of the cellular DNA polymerases, specific enzymes are required to digest the RNA moiety of the DNA · RNA hybrid. In bacteria this reaction may be catalyzed by a $5' \rightarrow 3'$ exonuclease activity of DNA polymerase I (*153*). Enzymes, for example, RNase H (EC 3.1.26.4), that digest the RNA part of DNA · RNA hybrids are present in higher organisms, but a second activity must accompany it to remove rN-dN linkages. So far, no RNase H enzymes are known to contain this activity (*162*).

DNA ligase, required for the joining of Okazaki fragments in prokaryotes (*153*), has been isolated from several mammalian species and characterized. This enzyme requires ATP for its function (*163–166*). It is quite likely that DNA ligase is involved in SV40 DNA replication. In fact, a DNA ligase was isolated earlier from mouse cells infected with polyoma virus (*164*).

Enzymes may be required for the unwinding of superhelical turns in DNA. Examples are the ω protein of *E. coli* and the "nicking–closing" enzymes isolated from rodent, human, and *Drosophila* cells (*167, 168*). Such enzymes appear as potential candidates for the "swivel" required during SV40 DNA replication. A part of the intracellular nicking–closing activity is associated not only with the SV40 chromatin (*21, 150*), but also with the viral DNA as a dodecyl-sulfate-resistant complex (*169*) and it mediates nucleosome assembly *in vitro* at physiological ionic strength (*170*), which is consistent with its presence in the SV40 replicating complex.

The involvement of an ATP-requiring "DNA gyrase" (*153*) to maintain the superhelicity of newly replicated DNA seems unlikely after the discovery that the superhelical turns in deproteinized SV40 DNA are introduced by the association of cellular histones as discussed earlier (Section II,A). But the role of DNA gyrase in the initiation of DNA replication cannot be ignored. Recently, an ATP-dependent "unknotting" activity has been detected in crude extracts of *Drosophila* embryos and *Xenopus laevis* eggs (*171*); it is thought to have a role in the initiation of DNA replication.

Among proteins that are involved in DNA replication are the "helix-destabilizing proteins" (HDP), such as the gene-32 product of bacteriophage T4 (*172*). Such proteins have been isolated recently from calf thymus (*173*). In the SV40 nucleoprotein complex, a single-strand-specific DNA-binding protein that stimulates DNA polymerase α activity has been detected (*150*). It is not known whether it interacts uniquely with the template and/or with the polymerase itself.

C. Assembly of Histones on the SV40 Genome

Replication of the eukaryotic chromosome is evidently different from that of naked DNA, as the former includes the coordinate synthesis of histones in the cytoplasm, their transportation to the nucleus, and their association with the newly replicated DNA. The SV40 genome to be replicated exists in the cell nuclei in the form of minichromosomes, not as naked DNA. It appears that, with the exception of the viral gene-A product, it employs the cellular enzymatic machinery for its replication. At the beginning of infection, the syntheses of both the cellular DNA and histones are stimulated (*174*), and these are tightly coupled (*175*), but later, cellular histone synthesis is uncoupled from cellular DNA replication and becomes coupled with viral DNA replication (*174*). The overall process of the assembly of histones with SV40 DNA resembles eukaryotic chromosome replication. Since the histones exhibit no pronounced turnover (*176*), they are passed between generations and, therefore, the central question remains: How do the parental and newly synthesized histones segregate both within the nucleosome and along the DNA strands in the course of DNA replication? For clarity, the nucleosomal structure is described below.

1. STRUCTURE OF NUCLEOSOMES

The repeat unit in chromatin (177), the nucleosome, appear under the electron microscope as particles about 110–125 Å in diameter (178). They contain a core of histone octamer, consisting of a pair of each of the histones H2A, H2B, H3, and H4, over which is wrapped about 200 base-pairs of DNA that varies form one organism to another (for reviews, 179–181). Probably one molecule of histone H1 is associated with each nucleosome. The packing ratio of DNA in a nucleosome is about 7. A limit digestion of native chromatin or of nucleosomes with microccocal nuclease produces a "core" particle that always contains about 140 base-pairs of DNA associated with the histone octamer but devoid of H1. The nucleosome core represents the basic level of organization common to all chromatin irrespective of the tissues or species. The additional DNA length present in the nucleosome (about 15–100 base-pairs) over that in the core particle is termed "linker DNA", and it is species- or tissue-specific. The linker DNA offers the binding site for H1 and other nonhistone proteins that play a key role in the higher order folding and in the function of chromatin.

It has been possible to elucidate some of the three-dimensional features of the nucleosome core by X-ray crystallography and electron microscopy (182). The nucleosome core is roughly a disk-shaped particle of diameter 110 Å and thickness 57 Å, somewhat wedge-shaped, and divided into two layers along its short axis. DNA is wound in a flat superhelix of pitch about 28 Å and average diameter about 90 Å outside the histone core. The wedge-shaped appearance suggests that there is less than two complete turns of DNA superhelix. Since the average diameter of the nucleosome core is about 90 Å, it appears to be consistent that the 140 base-pairs of DNA can make about 1.75 turns of superhelix with about 80 base-pairs per superhelical turn. Analysis of the digestion pattern of the core nucleosomes with DNase I also leads to the same conclusion (183).

2. SEGREGATION OF HISTONES IN NUCLEOSOMES

Experimental evidence under this heading comes mostly from studies of nucleosome assembly in cells rather than from the SV40 chromatin itself. Since the basic subunit structure of the SV40 minichromosome and its protein composition are the same as those found in higher organisms, the process of nucleosome assembly is likely to be the same in both cases. Three different models for this process are possible (*184*), namely, conservative, semiconservative, and random.

The segregation of histone is said to be conservative if the old and new histones are assembled into each octamer separately without mixing with each other. In the semiconservative mode, new octamer is thought to be made up of two heterotypic tetramers, one formed with old histones and the other with new histones. Any other distribution of histones within the new octamer is said to be random.

The assembly of new and parental histones into new nucleosomes during chromatin replication has been analyzed by equilibrium centrifugation in density gradients, using chicken myoblasts that were labeled with heavy amino acids and tritiated lysine (*184*). Under the experimental condition, about one histone molecule in eight was labeled. These histones were then fixed as octamers, and the DNA was removed with cesium formate and guanidinium chloride before equilibrium density centrifugation. From an analysis of the density of the labeled histone octamer, it appears that new histones are not mixed with the old ones inside a new histone octamer, i.e., the assembly is conservative (*184*), and that the old and new histone octamers segregate conservatively over two or three generations. It is still not known how histone H1 distributes itself in the new nucleosomes. These results exclude also the possibility of the semiconservative distribution of histones.

It was implicit in the original model (*177*) that a nucleosome, whose histone core is constituted of two each of the four histones, may open up into two separate half-nucleosomes, provided the nucleosome has a dyad axis of symmetry. During the last few years a number of models have been proposed describing the organization of DNA inside the nucleosome, two of which are based on two separate half-nucleosomes (*185, 186*). Electron microscopic evidence indicates that at very low ionic strength nucleosomes can be split up into two half-nucleosomes of diameter 90 Å each (*187*). Since semiconservative distribution of parental and newly synthesized histones is unlikely, it is not clear whether the half-nucleosomes seen under the electron microscope, have any function relevant to chromatin replication. On the other hand, the conservative segregation of histone octamers strongly suggests that the octamers do not leave the DNA during replication.

A recent study (*188*) indicates that newly synthesized histones do not associate synchronously as an octamer on the newly replicated DNA. Histones H3 and probably H4 are first assembled with the DNA, followed by the association of H2B and H2A. This result is consistent with those obtained *in vivo* with *Drosophila melanogaster* tissue culture cells (*189*), or in *in vitro* studies of core particle reconstitution (*189, 190*), and seems to rule out the semiconservative segregation of histones on the newly replicated DNA.

3. SEGREGATION OF HISTONES ALONG THE DNA STRAND

Models showing different possibilities of the segregation of histones along the DNA strand appear in a recent review (*191*). The replicating SV40 chromosome, like cellular chromatin (*192*), contains nucleosomes in both arms that are indistinguishable from one another under the electron microscope (*193, 194*). Pulse-labeling indicates that newly synthesized histones associate preferentially with the replicating intermediates, but it was not clear whether the newly synthesized histones segregate with the unreplicated portion of the genome or with the replicative "eye" (*193*). The segregation of parental histones was studied in the absence of protein synthesis by incubating virus-infected cells in puromycin for 1 hour prior to labeling replicating intermediates. Such complexes sediment at about 25 S and contain about half the number of nucleosomes normally present in the minichromosome. The number of base-pairs in each nucleosome is about 125, much less than the native nucleosome (*193*). Furthermore, about 30% of the nucleosomes were spaced at distances similar to those of the mature minichromosome, while the remainder were at greater distances. Interestingly, no naked DNA was observed in the whole population. These results were interpreted by the authors as an indication of the random distribution of parental nucleosomes, but are subject to the obvious criticisms that (*a*) the cells were kept in the inhibited condition for more than three to four times their generation time; (*b*) no structure contained a replication eye; (*c*) since SV40 DNA replication is bidirectional, the absence of the naked DNA does not necessarily mean that the conservative segregation of nucleosomes cannot occur. However, there are reports from other laboratories that indicate a random distribution of nucleosomes (*195–197*).

On the other hand, in several systems in the absence of protein synthesis, using micrococcal nuclease and DNase I as probes, newly synthesized DNA was twice as sensitive to nucleases as was control chromatin (*198–200*). Partial nuclease digestion of the newly replicated DNA with micrococcal nuclease revealed a normal pattern of monomer and multimer (*198, 200*). These results indicate that newly synthesized chromatin in the absence of protein synthesis contains only half the number of nucleosomes of parental chromosomes, and these are aligned cooperatively in a conservative fashion on only one of the daughter strands, yielding a normal digestion pattern. Electron-microscopic observation (199) of long stretches of protein-free, non-beaded DNA in the replicating chromosome in the absence of protein synthesis also supports the idea of conservative segregation of histones.

On the basis of the basic asymmetry of replication forks and the conservative segregation of parental histones, it is proposed (199, 201) that the leading side of the replication fork, where the DNA synthesis is continuous, is covered by parental histones, whereas the lagging side receives the newly synthesized histones. The same parental strands at the leading sides of both replication forks serve as templates for the coding of early and late viral mRNAs. These results are valid also for chicken cells in culture, where parental histones segregate asymmetrically and are associated with those DNA template strands that code for stable nuclear RNA species detected by hybridization to single-copy DNA (201). This asymmetric segregation of parental histones to one of the daughter strands may provide a vehicle for transmitting and segregating chromosomal information during development. The association of Okazaki fragments with nucleosomes was demonstrated in CHO cells and HeLa cell nuclei (202, 203). In contrast to these reports, it has been shown that, in SV40 replication, Okazaki fragments are not contained within the nucleosome (117).

An "assembly factor," purified from cell extracts of *Xenopus laevis* and identified as an acidic protein of molecular weight 29,000, plays a role *in vitro* in the transfer of histones to the nucleosomal sites, probably by neutralizing the positive charges of histones and thus reducing their nonspecific association with SV40 DNA (204). Although not purified, such an activity is also present in other eukaryotic systems, including *Drosophila* (205) and the host BSC-1 cells for SV40 (206). However, it appears that some highly acidic proteins, like poly(glutamic acid) or poly(aspartic acid), also assemble histones into an octamer, thus facilitating the process of nucleosome formation in agreement with the acidic nature of the purified assembly factor (207). As discussed earlier, a nicking–closing enzyme also seems to mediate the process of nucleosome assembly (170).

IV. Transcription of the SV40 Genome

In this section, we concentrate on the transcription of the SV40 genome during lytic infection of permissive monkey cells, rather than during abortive infection of nonpermissive cells. The latter is discussed in a recent review (10), where the sequences of the transcription units and their control have been extensively discussed, and will not be repeated here.

The lytic cycle is divided into two phases. The early phase begins with the uncoating of the virus and lasts until the beginning of viral DNA replication, which occurs at about 12–20 hours after infection at 37°C. Virus-specific antigens T and t are synthesized in this period. The late phase begins with the commencement of viral DNA replication and terminates at cell death. In this phase, viral capsid proteins are synthesized and progeny virions are assembled in the nucleus of the cell.

Escherichia coli RNA polymerase can transcribe the entire length of SV40 DNA asymmetrically (*208, 209*), which made it possible to use the RNA synthesized *in vitro* (SV40 cRNA) as a direct probe for virus-specific transcripts of complementary sequences and also as a tool to separate the strands of SV40 DNA to use as probes for RNA synthesized *in vivo*. Both approaches indicate that early viral RNA had the same polarity as cRNA and was copied from the minus (−) strand of SV40 DNA, whereas late RNA consists of two types of transcripts, one indistinguishable from early RNA and the other complementary to cRNA, and copied from the viral plus (+) strand (*210–213*). The direction of transcription of the SV40 genome was determined either by determining the 3′ → 5′ polarity of the SV40 minus strand (*214*) or by locating the specific cRNA initiation site on the restriction fragments of the SV40 genome (*215–217*). The SV40 minus strand, coding for cRNA and early mRNA, was transcribed in the counterclockwise direction whereas the (+) strand was transcribed in the clockwise direction in the conventional map.

A. Transcription of SV40 Genome *in Vivo*

1. EARLY TRANSCRIPTION

Early virus-specific RNA represents only about 0.01% of the total RNA synthesized in infected cells (*218–220*). Hence, it is not possible to label early nuclear RNA for periods of less than 30 minutes, whereas late RNAs can be pulse-labeled in a period as short as 5 minutes. Many of the earlier data on the size of the virus-specific early primary transcripts are difficult to reconcile because of the different systems and labeling conditions used by different authors.

Although some virus-specific RNA sediments heterogeneously in the range of 20–26 S or 30–50 S, it appears that the 19 S component represents the size of the major primary transcripts and corresponds to about half the size of the viral (−) strand (221, 222). Primary early transcripts, isolated from cells infected with SV40 tsA mutants, hybridize to about 70% of the viral genome, whereas a small amount of nuclear RNA spans less than 20% of the genome (223). The relevance of this finding to the size of the early primary transcript is questioned because the experiment was done with tsA mutants. The primary transcript in this system has its 5′ terminus at approximately 0.66 map unit as observed under the electron microscope (223), but the 5′ terminus for the wild-type virus is not yet known.

Although most of the early nuclear transcripts are derived from the early region of the SV40 genome, hybridization data indicate the presence of transcripts from all or a portion of anti-late regions of the viral (−) strand, as well as from all or a portion of the (+) strand in the nuclei of infected cells early in lytic infection (224, 225). While observations contrary to these findings exist (226, 227), further studies with early and late transcription complexes isolated from cells infected with wild-type SV40 or its tsA mutants indicate that RNAs complementary to early and anti-late regions of the (−) strand, and late and anti-early region of (+) strand, can be synthesized *in vitro* (228–230). In polyoma-infected cells, virus-specific nuclear RNAs also hybridize with either strand of DNA fragments both within and beyond the early region (231). A technique sensitive enough to detect even less than one RNA molecule per cell shows that, in the cytoplasm of monkey cells infected with wild-type virus or DNA, much smaller amounts of two late mRNAs together with the early mRNA can be detected in the cytoplasm (232). All these results taken together indicate that both strands of SV40 DNA are transcribed extensively early in lytic infection, but the major stable primary nuclear transcript is the 19 S RNA that contains mainly early gene sequences.

Cytoplasmic early virus-specific mRNA (about 0.01% of the total cytoplasmic RNA) consists of one component of 19 S that hybridizes to about 50% of the (−) strand (213, 225). This mRNA has a poly(A) stretch about 150–200 nucleotides in length at its 3′ end (233), but there is still no direct evidence for the presence of a "cap" structure at the 5′ terminus, or of methylation of internal bases. From hybridization studies with the separated strands of different restriction fragments, it was concluded that the 5′ end of the early mRNA lies near the origin of DNA replication at about 0.66 map unit, and the 3′ end at about 0.15 map unit (119, 216, 224, 225, 234, 235). Early mRNA is the result of posttranscriptional processing of the primary transcript either in the nucleus or in the cytoplasm and consists of two different size classes of mRNA that are translated into two different proteins (118,

119, 236). The sedimentation profiles indicate that the early mRNAs do consist of two different populations (*120, 237*).

2. LATE TRANSCRIPTION

The late primary transcripts of virus-specific RNAs represent about 10% of the RNA labeled in infected cells (*238*) and sediment in sucrose gradients with sharp peaks at 19 S, 23–24 S, and 26 S (*239–243*). A broad profile extending from 27 to 60 S is also occasionally observed (*221, 244*). RNAs of 19–26 S usually constitute 80–100% of the total viral RNAs, while species of higher molecular weight make up 0–20%. Some of the pulse-labeled late primary transcripts are self-complementary, and therefore can form double-stranded RNA · RNA hybrids, indicating that both (+) and (−) strands are transcribed (*239, 245*). Hybridization of the unlabeled nuclear RNAs to the labeled separated strands of SV40 DNA, or restriction fragments thereof, clearly demonstrate that all (+) strand and all or most (−) strand sequences are present in late nuclear RNAs (*224, 225, 239, 245*). The (+) strand transcripts are 10- to 20-fold more abundant than the (−) strand transcripts (*225, 243*); this may arise from more efficient transcription rather than from higher metabolic stability (*241*). It has also been shown that the transcript of the late region of the (+) strand is about 3- to 5-fold more abundant than the anti-early region of this DNA strand (*246*).

Nuclear RNAs both with or without poly(A) have been detected and analyzed (*247*). The nuclear RNA-poly(A) contains at least five different species not found in the cytoplasm, with the same 3′ end (at 0.17 map unit) as that found on cytoplasmic viral RNAs, but with a different heterogeneous position of the 5′ end located at 0.72, 0.70, 0.67, 0.64, and 0.59 map units. The RNA species that maps from 0.72 to 0.17 is the most abundant nuclear RNA, and the different species, which appear to be a colinear subset of one another, most probably represent intermediates of a stepwise progression in the posttranscriptional processing of the 5′ ends. The 5′ ends of the nuclear RNA without poly(A) are the same as the poly(A)-containing nuclear RNA, the relative abundance being the same, but the 3′ ends of these RNA molecules map at 0.20, 0.21, and heterogeneously beyond 0.28 (*248, 249*). Based on this observation, it has been proposed (*247*) that longer non-polyadenylylated viral RNA molecules in the nuclei of SV40-infected cells are the primary late transcripts. The heterogeneity found in the leader segments may indicate that a few nucleotides on the 5′ end of some 16 S or 19 S RNAs are derived from sites proximal to 0.72 map unit or that the transcription of late RNA is initiated at multiple sites.

Late mRNA generally accounts for 1–3% of the total cytoplasmic

RNA labeled in several hours (244, 249). It sediments in sucrose gradients in two peaks with sedimentation coefficients of 16 S and 19 S. The 19 S class includes a low level of the early RNA species and a preponderance (~95%) of the late RNA species. This early RNA species is complementary to the same half of the (−) strand that codes for the 19 S RNA appearing at early times, while the late 16 S and 19 S mRNA species are complementary to the opposite half of the (+) strand (210, 213, 225). The late RNA species have been mapped by hybridization with SV40 Hind fragments and shown to be transcribed clockwise from 0.65 to 0.17 map unit (215, 225, 233). The 16 S species is about 10 times more abundant than the 19 S species and both have the same 3′ terminal (216, 235, 250). Neither 16 S nor 19 S late RNAs are coded from contiguous segments in the region of 0.65 to 0.17 map unit. Each RNA contains a leader sequence of about 150–200 nucleotides at its 5′ end, joined to a main body of coding sequences, and not transcribed from a continuous sequence of the viral DNA (251–254). Both the 16 S and 19 S late mRNAs consist of a few different subspecies having different 5′-terminal leader sequences (247). All the mRNA species are polyadenylylated, methylated, and capped at their 5′ end (255–258). However, as many as eight distinct 5′-terminal capped structures have been detected in SV40 late mRNAs (259).

B. Transcriptional Control of Gene Expression

It appears from the previous discussion that control mechanisms operate in both early and late gene expression. At the early stage of infection, only one half of the (−) strand is predominantly transcribed whereas in the late stage the (+) strand of the other half of the duplex is transcribed into stable mRNA. Minor quantities of late mRNAs at early times, and early mRNAs at late times, after infection are also present. The expression of the late genes is dependent on the early gene-A product, the T antigen. The crucial role played by T antigen is not yet understood. Interestingly, both the 5′ ends of early and late mRNAs are situated at or near the origin of replication. Whether this reflects a relationship between control of transcription and replication is not known.

The aim of this section is to analyze the signals for promoters and terminators, and the transcriptional control mechanisms, at three different levels, namely, RNA polymerase, template, and gene products, that may be involved in the regulation of gene expression.

1. Promoter and Terminator Signals

Different lines of evidence suggest that the promoter for early transcription lies within the origin of replication and about 60 nucleotides upstream from the template for the 5' terminus of early mRNA (260–262). From an analogy of certain bacterial and eukaryotic promoters, it appears that the promoter for early transcription lies at least in part within 0.66–0.67 map unit (10). It is believed that the (−) strand transcription late in infection is under the same control mechanism as in early transcription. The location of late promoters is also not definitively known due to the heterogeneity of the 5' end of late RNAs but is believed to lie on the (+) strand in the region of 0.58 to 0.72 map unit. Electron microscopic observation of nascent RNA chains on the late viral transcription complex, and the presence of late nuclear RNAs with 5' ends at positions 0.64 and 0.59 strongly suggest one or more initiation sites upstream from 0.67 (247, 263). Analysis of the genomic sequence also reveals that the origin of replication as well as the immediate upstream region contains a number of (A + T)-rich stretches (3, 4) that may serve as promoter signals, like prokaryotic systems (264, 265).

The potential signals for termination of early and late transcription have not yet been identified. Some of the evidence is derived indirectly from the analogy of bacterial systems and from the location of the promoters and size of the primary transcripts. Out of the 20 stretches of six or more dA residues on the SV40 genome that may act as potential termination signals, only the three located at 0.654, 0.666, and 0.155 map units are good candidates (10). Only the dA residues at 0.654 map unit, which is preceded by a relatively (G + C)-rich region within the origin of replication and about 30 nucleotides downstream from the position of 5' termini of early mRNAs, may act strongly as an effective terminator (264, 265) for full-length (−) strand transcript. The 3' ends of the (+) strand transcripts are also heterogeneously distributed (247, 248). From comparisons of plus strand nucleotide sequences with that of the known bacterial termination signal (264) it appears that two dA residues at 0.739 and 0.180 map unit may act as termination signals for unit and for half genomic length RNAs, respectively.

2. RNA Polymerase and Other Protein Factors

Although it is possible that early and late transcriptions may be carried out by two different RNA polymerases, a new RNA polymerase being synthesized late in infection, or the one engaged in early transcription being modified so that it can now recognize and transcribe the late viral DNA sequences, compelling evidence for such is not available. Synthesis of both early and late virus-specific RNAs in isolated nuclei or with late transcription complexes is very sensitive to α-amanitin (266, 267), which inhibits especially the mammalian RNA polymerase II (160). It thus appears that the host RNA polymerase II is involved in the specific transcription of both (−) and (+) strands of SV40 genome.

It was shown earlier that both mammalian RNA polymerases I and II can transcribe superhelical SV40 DNA *in vitro* from both strands and from all regions of the DNA, indicating that no selective transcription of specific sequences occurs (268, 269). *Escherichia coli* RNA polymerase, on the other hand, is able to transcribe selectively the (−) strand of SV40 DNA (208), although none of the five identified binding sites corresponds to 5' termini of *in vivo* viral early mRNAs (217). It is not clear what relevance these results have to *in vivo* transcription. Specific factors may be missing in the purified mammalian RNA polymerase II, or in the purified DNA template that may be responsible for transcriptional specificity. In fact, until very recently, specific transcription *in vitro* by class II RNA polymerases that transcribe class II genes (which encode mRNAs) *in vivo* has been a major challenge in eukaryotic molecular biology. Selective and accurate initiation of transcription has recently been achieved using adenovirus DNA (270, 271) and cloned conalbumin and ovalbumin genes (271) with either homologous or heterologous RNA polymerase II preparations, and in the presence of S100 extracts of mammalian cells that confer specificity. An alternative mechanism of specific initiation by the introduction of specific nick(s) in SV40 DNA has been recently suggested (272, 273). The presence of an endogenous endonuclease in SV40 chromatin that recognizes the region between 0.67 and 0.73 map units, which is also the region for both early and late promoters, is suggestive in this regard.

A core SV40 nucleoprotein complex, isolated by mild disruption of mature virions and containing all three capsid proteins, can be transcribed more efficiently *in vitro* by both *E. coli* RNA polymerase and eukaryotic RNA polymerase II (35, 274), compared to relaxed SV40 minichromosome that contains only VP3 in addition to cellular histones. RNA synthesized by *E. coli* RNA polymerase sediments as a fairly homogeneous species of about 16–18 S, and the transcription is asymmetric, occurring preferentially on the (−) strand of the SV40 genome (35). In contrast, transcription with RNA polymerase II is symmetric. These results indicate that viral capsid proteins may have a role in the regulation of early gene expression. A similar conclusion was reached independently with SV40 transcriptional complexes isolated from cells infected with deletion mutants within VP2–VP3 gene (G. R. Stark, personal communication).

3. TEMPLATE

Form-I SV40 DNA acts as the template for early transcription *in vivo*. An alteration in the physical structure of the viral genome may be responsible for the change in the transcription pattern late in infection. This alteration may include the removal or addition of histones and/or other acidic proteins acting as repressors, coupled with the initiation of viral DNA replication to prepare the template for late transcription. It is also possible that some specific structural changes of the template may occur in its interaction with the early gene product before the onset of late transcription.

a. Does Replicating DNA Act as Template for Late Transcription? Initiation of DNA replication is required for the expression of the late genes of SV40. The early gene-A product—the T antigen—is required for the onset of DNA replication (76, 77) and is known to bind to a stretch of DNA of about 120 base-pairs containing the origin of replication and possibly also the early and late promoters (135, 137). It is controversial whether T antigen directly regulates both the early and late gene expression, or whether replicating DNA is the template for the latter. Girard *et al.* (275) isolated from infected cells a transcription complex consisting of relatively short chains of SV40 specific RNAs hydrogen-bonded across a fraction of their lengths to SV40 DNA, and suggested that the DNA may be in the form of a replicative intermediate. In a more recent study (276), replicative intermediates of SV40 DNA (wild type or *ts*A mutant) were micro-injected into different cell lines, including CV-1, that were blocked in DNA synthesis at the restrictive temperature. Late gene expression was obtained in a high percentage of recipient cells either at 37°C or 41.5°C, whereas form-I DNA at a 10-fold higher concentration failed to induce V antigen (capsid proteins) synthesis. However, in this study, V antigen synthesis was also obtained by micro-injection of partially denatured SV40 DNA or randomly nicked SV40 in the absence of DNA synthesis. Thus, it is not clear whether late transcription needs a single-stranded region or a single nick at or near the origin of replication. The role of T antigen in this crucial structural change is not yet understood.

In studies with *ts*A mutants or in the presence of inhibitors of DNA synthesis, form-I, and not replicating, SV40 DNA appears to be the template for late transcription. Two types of approach were made. First, in cells infected with *ts*A mutants at restrictive temperatures, no late mRNA was produced, but when these were infected at permissive temperatures and then shifted to restricted temperatures such that no new cycle of DNA replication was initiated, late mRNA synthesis continued (272). Studies *in vitro* with both early and late viral transcriptional complexes (229, 277, 278) isolated from cells infected with either wild-type or *ts*A mutant lead to the same conclusion, namely, that replicating intermediates do not serve as templates for late transcription; rather, viral T antigen seems to be directly involved. Second, in the presence of inhibitors of DNA synthesis, such as arabinosylcytosine (araC), late mRNA is always detected, although in reduced amounts (232). Both approaches strongly indicate that continued DNA synthesis is not essential for late transcription.

When late viral transcription complexes are examined under the electron microscope, RNA chains are found on both relaxed and superhelical molecules as well as oligomers of SV40 DNA, but not on replicating molecules (263). This suggests that once a molecule starts to replicate it cannot serve as template for transcription until the replication is complete. It has also been shown that "recently" replicated SV40 DNA is utilized preferentially as template for both transcription and replication (279). However, in the presence of arabinosylcytidine, DNA synthesis may be initiated, but there should be no complete rounds. It therefore appears unlikely that new, fully replicated DNA is the only template for late transcription. One should be cautious in interpreting work done with different *ts*A mutants, since it appears that the expression of late genes depends on the cell line (280). In AGMK cells, an early gene product is continuously required for efficient expression of late viral genes. In contrast, late gene expression, once initiated in CV-1 cells, continues efficiently regardless of *ts*A mutation.

b. Minichromosome as Template for Transcription. Since attempts to reconstitute an *in vitro* system for specific transcription from purified RNA polymerase and viral templates failed, the problem was approached with different kinds of viral transcriptional complexes (VTC) isolated from infected cells. One of these complexes is soluble in Sarkosyl (281) and the other in Triton (282), but both contain short RNA chains already initiated *in vivo*. With these complexes, elongation of RNA chains and their specific termination have been studied *in vitro* with radioactive precursors. The Sarkosyl-soluble one is about 25 S and consists mainly of DNA, RNA polymerase, and RNA (278, 281, 283). In contrast, the Triton-soluble VTC contains most of the cellular histones, in addition to the RNA and RNA polymerase, bound to the DNA template; the sedimentation coefficient of the complex 50–55 S, is similar to one form of the SV40 minichromosome (282, 284, 285). A recent study involving electron microscopy, density measurement, and sedimentation behavior indicates that the latter VTC has a structure indistinguishable from that of a native SV40 minichromosome (286).

When the Sarkosyl-soluble VTC is transcribed *in vitro*, about 5% of the RNA is complementary to the E-strand and 95% to the L-strand (266, 283). Syntheses of both RNAs are inhibited to the same extent by low concentrations of α-amanitin, indicating that both strands are transcribed by RNA polymerase II. About 70–90% of RNA elongated *in vitro* is virus-specific and synthesized from all of the E- or L-strand. Although a low level of reinitiation of RNA synthesis occurs, none of this reinitiated RNA is virus-specific (283). A similar conclusion was reached with VTC isolated from polyoma virus (287).

With Triton-soluble VTC as template, the nucleosomal DNA is available for transcription and the *in vitro* synthesized transcripts range from 10 S to 30 S (on prolonged incubation) (284, 286). Since no new initiation of RNA chains occurs under the incubation conditions, the *in vitro* product is elongated from RNA chains initiated *in vivo*. The sizes of the RNA synthesized on prolonged incubation appear to be much longer than the late transcripts *in vivo*, indicating that no specific termination occurs. Most of the transcriptional complex contains one RNA chain, in contrast to 10 RNA chains observed in Sarkosyl-extracted VTC (263). However, titration of SV40-transcribing RNA polymerase II molecules with [^{3}H]α-amanitin indicates that there are about two RNA polymerase II molecules per SV40 transcriptional complex, in agreement with electron microscopic observations (286). Ammonium sulfate increases the rate of transcription by direct action on the enzyme and also by causing a reversible change in the viral chromatin, especially at the nucleosomal level (284, 286).

In several cases, viral minichromosomes isolated from the mature virions (288–290) and nucleoprotein complexes reconstituted with SV40 DNA and cellular histones (291, 292) have been used as RNA-free templates in *in vitro* transcription studies with both eukaryotic and prokaryotic RNA polymerases. It appears that the nucleosomal DNA is available for transcription in both cases (288–292), in agreement with results obtained with Triton-soluble VTC (284–286), and that the physical state of the minichromosome is not altered as a result of transcription (288, 289). The RNA transcripts synthesized in a 30-minute incubation with *E. coli* RNA polymerase sediment as 3 S–10 S particles but the transcript of a KCl-treated sample is much longer, about 16 S (288), indicating that the electrostatic interactions between the histone octamer and DNA must be broken to facilitate the transcription of nucleosomal DNA. From these studies, it appears that elongation of the RNA chains and selection of the DNA strand by *E.*

coli RNA polymerase are greatly reduced in the presence of nucleosomes (*288, 290*). No region of the SV40 minichromosome is preferentially transcribed (*290*) either by *E. coli* RNA polymerase or by RNA polymerase II. Recently, it has been shown that SV40 minichromosomes, isolated late after infection by a low-salt detergent-free extraction procedure (*52*), can be transcribed by *E. coli* RNA polymerase along the late strand in a clockwise direction (E. B. Jakobovits, S. Sragosti, M. Yaniv, and Y. Aloni, personal communication). The synthesis is initiated within a DNA fragment spanning 0.67–0.76 map units on the genome that contains no nucleosome. When compared with the transcript obtained with naked DNA, this result strongly suggests that the nature of the template alters the transcriptional specificity.

4. ROLE OF T ANTIGEN

In a lytic infection, gene A is transcribed before the onset of viral DNA replication, thereby leading to the appearance of early mRNA in the cytoplasm and its translation into the T antigen (*118, 293*). Temperature-sensitive mutants in gene A produce a more thermolabile T antigen (*293, 294*) that fails to function in the initiation of DNA synthesis at a nonpermissive temperature, indicating its positive role in DNA synthesis (*76*). The T antigen binds preferentially to the origin of replication *in vitro* (*135, 137*) although it can also bind to the other regions of SV40 DNA (*133*).

In infected cells, T antigen is synthesized in an exponential fashion up to about 96 hours, and accumulates in the cell nucleus (*293, 294*). Under restrictive conditions, although the steady-state amount of the T antigen is very low in cells infected with *ts*A mutants (*295*), the rate of synthesis of this protein is more rapid than in a corresponding infection with wild-type virus (*294*). This strongly suggests that T antigen regulates the expression of early genes, i.e., its own synthesis; the faster synthesis may reflect the loss of a negative control in the presence of a thermolabile T antigen. This idea was later confirmed and extended to show (*127, 296, 297*) that *ts*A mutants (*a*) overproduce early RNA, and this occurs at the level of initiation of transcription, even at permissive temperatures, and the effect is amplified greatly with a shift to nonpermissive temperature; and (*b*) have a diminished ability to synthesize viral DNA at both permissive and restrictive temperatures.

These properties of the T antigen, taken together, indicate its dual function in the initition of viral DNA synthesis and in the negative feedback regulation of its own synthesis (*296*). Since the 5' ends of RNA molecules complementary to the early strand of SV40 DNA map near the origin of replication, the control of early transcription may well be achieved by the interaction of the T antigen with a stretch of nucleotide sequences containing the origin of replication and both early and late promoters (250). During early infection, production of large amounts of early RNA by tsA mutants, especially at 41°C, indicates that the early promoter is very efficient (294, 296). This offers the virus the additional advantage of being able to initiate early transcription and the processes that follow and depend on it. However, further control requires diverting the genome from early transcription to replication and late transcription (232).

Late transcription begins after the expression of the A gene with concomitant initiation of DNA synthesis (210, 211, 213). Two possible modes of controlling late gene expression are apparent. In the first mode, a protein repressor associated with viral DNA in the virion blocks initiation or progress of RNA polymerase on the late DNA strand until early transcription or DNA replication generates a repressor-free template (232). This idea of negative control by a repressor protein stems from the observation that tsD mutants fail to synthesize early RNA (298) whereas protein-free DNA from this mutant can initiate infection normally, probably owing to the prior removal of the repressor from the template (299, 300). In the second mode, T antigen may be involved either directly in late transcription or indirectly in the sense that it initiates DNA replication, and DNA replication per se is needed in the initiation of late transcription (232).

If the second model is valid, infection with tsA mutants or DNA thereof at the restrictive temperature should fail to synthesize late mRNA. Results obtained earlier with the tsA mutant indicate that the function of A gene is required for transcription of late viral genes but once initiated, synthesis of T antigen is no longer required for continued late transcription (126, 229, 277). In another study (232), infections with tsA 58 virus or DNA at the restrictive temperature led to substantial amounts of two early RNAs, but no late RNAs, whereas in cells infected with wild-type virus or its DNA, a significant amount of late mRNAs was synthesized 7.5 hours after infection. These results, which agree with those observed with early transcription complexes isolated from tsA 58-infected cells (229), strongly suggest that the T antigen plays an active role in late transcription and are contrary to the idea of negative control by a repressor, unless the repressor is resistant to vigorous deproteinization or is a host molecule that binds to SV40 DNA when the virus is uncoated.

It has been noted that when infection with tsA 58 virus or its DNA was prolonged for about 24–30 hours at restricted temperatures, both forms of late RNAs were detected (232, 297), although most of the antigenic reactivity of T antigen is lost rapidly at 41°C. Since neither tsA 58 virus nor tsA 58 DNA form plaques at 41°C, it is unlikely that these late RNAs could have arisen from a few-wild type revertants in each stock. As pointed out earlier, it is possible that initiation of DNA replication requires large T antigen at a much higher concentration than does late transcription. The low steady-state level of functional T antigen that gradually accumulates in a tsA mutant-infected cell at the restrictive temperature (297) is insufficient for DNA synthesis but sufficient for late transcription. Large T antigen may stimulate late transcription when bound at or near the origin (134, 135, 137) at low concentrations and, additionally, may stimulate initiation of viral DNA replication and repress early transcription at higher concentrations (127).

It is still not clear whether the regulation of late gene expression by T antigen is distinct from its role in initiating DNA replication or in repressing early transcription. Late transcriptional complexes, with short nascent RNA chains initiated near the origin of replication, accumulate late in infection (263). This suggests that late transcription may be blocked at a nearby attenuator site until a stimulatory factor, which may be the T antigen, allows transcription by stoichiometric binding to the DNA or by catalytic removal of an attenuator protein. This idea appears consistent with the results of strand-specificity of SV40 RNA synthesized from transcriptional complexes isolated very early in infection with either Sarkosyl or Triton X-100 and using either wild-type or tsA 58 virus (229). About 20% of the RNA synthesized *in vitro* with this complex was late RNA, strongly suggesting the removal of a postinitiation block to late transcription during the isolation of the complexes in the presence of detergents.

C. Posttranscriptional Control

The fact that both early and late mRNA species are different from the primary transcripts in regard to size and composition points to the involvement of processing mechanism(s) operative at a level subsequent to transcription. This processing includes the addition of poly(A) at the 3' end, capping at the 5' end, methylation of bases, and splicing of intervening sequences (IVS) of the primary transcript (13–15). While the last one determines which sequences will be exported to the cytoplasm as mRNA, the 5' cap and 3' poly(A) structures are important for ribosome binding, mRNA translation, and protection against nucleolytic degradation (13, 14). The sequences that are spliced out are termed "introns" and those joined together are called "exons" (*301, 302*). In the following sections, mainly the splicing of early and late SV40 mRNAs will be discussed; for the molecular mechanism of other processing steps, the readers are referred to two review articles that extensively cover this area (*13, 14*).

1. PROCESSING OF EARLY mRNAs

Early in infection, two spliced viral mRNAs are detected in the cytoplasm of the host cell (*119, 236*). One of these is about 2200 nucleotides in length and is composed of two parts mapping from 0.67 to 0.60 and from 0.533 to 0.14, respectively, on the physical map of SV40. The other mRNA is about 2500 nucleotides long and is also composed of two parts coded from 0.670 to 0.546 and from 0.533 to 0.140 map units, respectively (*119, 236, 303, 304*). Translation of the 2500-nucleotide mRNA yields small t antigen whereas translation of the 2200-nucleotide one produces the large T antigen. Studies (*303, 304*) using primer extension methods with reverse transcriptase show that the mRNA for small t contains a segment of 588 bases (residues 18 to 605) spliced to residue 672, a 66-nucleotide segment from 0.546 to 0.533 map unit, rich in A and T being spliced out of this mRNA. The mRNA for large T contains a segment of 308 bases in length (residues 18 to 325) which is also spliced to residue 672. A segment of 346 base-pairs extending from 0.60 to 0.533 map unit is spliced out of this RNA (*304*). One of these studies (*303*) raised the possibility of an additional 5' terminus, as indicated by two principal reverse transcriptase stops, both occurring with equal frequency at the 5' termini of early SV40 mRNA. This has been ruled out by another study (*304*) that indicates that the two early mRNAs have identical 5' ends of sequences AUU located at residue 18–20. The DNA sequences between 0.67 and 0.60, present in both mRNAs, are translated in the same reading frame to yield two separable gene products that have the same NH_2-terminal sequences (*119, 236, 303, 304*).

Deletion at the distal position of the early region alters the structure of large T, but not of small t, whereas deletions within the region of 0.59 to 0.55 result in an alteration or absence of small t but yield a normal large T (*119, 305, 306*). Interestingly, the in-phase termination codon that stops small t translation has been spliced out in large T translation. Results of *in vitro* translation of SV40 cRNA (product of *E. coli* RNA polymerase) and SV40 early mRNA are consistent with this model (*118, 307, 308*). The sequences of the spliced junctions of the early mRNA for large T and small t are AACUC/AG/AUUCCA and CUAUA/AG/AUUCCA, respectively (*303, 304, 309*).

The other events of processing, involving the addition of poly(A) at the 3′ ends, is a posttranscriptional phenomenon, but precedes splicing (233). As mentioned earlier, it is believed that early mRNAs contain 5′-terminal cap structures and methylated bases, like late SV40 mRNAs or other eukaryotic mRNAs (257, 258). Since the two early mRNAs have identical 5′ termini and different spliced junctions, it appears that the expression of early SV40 genes is controlled both at the level of initiation of transcription of a single promoter and at the posttranscriptional splicing of mRNA to control the abundance of viral gene products.

2. PROCESSING OF LATE mRNAs

By hybridization experiments and electron microscopy the coding sequences for 16 S and 19 S late cytoplasmic mRNAs have been located between 0.935 and 0.17, and 0.765 and 0.17, map units, respectively (247, 251). The 5′ terminal 100–200 ribonucleotides of late SV40 mRNAs are not transcribed adjacent to their coding sequences (310, 311). The 16 S mRNA that codes for VP1 contains a leader sequence of approximately 200 nucleotides, corresponding to 0.72–0.76 map units, spliced to the coding sequences at the 0.935 map unit (247, 312). There are several different leader segments for the late 19 S mRNA that codes for structural proteins VP2 and VP3, each spliced to the same coding sequences. The sizes of these 19 S leader sequences are about 50–70, 100–120, 200–210 nucleotides. The 5′ ends of these leaders are located at 0.72, 0.71, 0.695, and 0.69 map units (247). With a viable deletion mutant dl 1811, a variety of capped structures was found at 5′ termini, indicating also a heterogeneity in the 5′ terminus of late mRNAs (313).

It was suggested that the 16 S RNA is processed from the 19 S RNA in the cytoplasm (233, 240). An analysis of the pulse-labeled viral RNA at 45 hours after infection (314) reveals that the nuclear viral RNA decays in a multiphasic manner leading to products having S values of 19, 17.5, and 16, respectively. Kinetic data indicate that the 19 S RNA is transported as such directly from the nucleus, and this RNA is the predominant cytoplasmic species seen in the short chase. The 16 S RNA is the predominant species in the longer chase and seems to be the spliced product of the 19 S RNA (314). The half-lives of the cytoplasmic 19 S and 16 S RNAs are about 2 hours and 5 hours, respectively. With nuclei isolated from infected CV-1 cells under mild isotonic conditions, it was demonstrated that the 19 S mRNA is the precursor to 16 S mRNA and that the splicing occurs in the nucleus (315).

Both late mRNA species are polyadenylylated at the 3′ end, methylated and capped at the 5' end (255–259). Since the 3' end of all poly(A)-containing nuclear RNAs is located about 0.17 map unit in contrast to the 3' end of non-polyadenylylated RNAs, which is located at 0.20, 0.21, and beyond 0.28 (247, 248), it appears that poly(A) addition to RNA molecules occurs after a specific cleavage of the longer transcripts. While the 5' ends are being processed, the specific addition of poly(A) takes place at 0.17 map unit. Polyadenylylation is then probably followed by splicing to generate the cytoplasmic mRNA. The capped structure at the 5' end consists of a 7-methylguanine (m^7G) residue linked 5' to 5' to the first transcribed nucleotide of an RNA chain by a pyrophosphate group. The principal 5' terminal sequences of both 16 and 19 S mRNAs are $m^7G(5')pppm^6Am$-U (47%), $m^7G(5')pppm^6Am$-Um-U (19%) and $m^7G(5')pppm^6Am$-C (15%) (257–259).

3. MECHANISM OF SPLICING

Hybridization of the precursor poly(A)-containing nuclear viral RNA with the "late" portion of SV40 DNA (0.67–0.17 map unit), including the corresponding interruptions between the leaders and the bodies of the cytoplasmic mRNAs, clearly indicates that splicing is a posttranscriptional process following polyadenylylation (263). However, it is not known definitively where splicing occurs. It is thought that it takes place either in the nucleus or immediately after the RNA is transported to the cytoplasm (240, 314, 315). No enzymes have yet been isolated that remove the IVS from the mRNA precursor *in vitro*. Recently, however, splicing of the 19 S mRNA to 16 S mRNA has been carried out in isolated nuclei (315).

Although the exact mechanism of splicing is not understood, four primary mechanisms by which the leader could be joined to the coding sequences have been postulated (251, 263): (a) deletion of intervening DNA sequences; (b) intermolecular ligation of RNA; (c) looping out of intervening DNA sequences so that RNA polymerase skips over short distances and continues; and (d) looping out of the intervening RNA sequences, excision, and rejoining. When viral transcriptional complexes with initiated RNA chains are observed under the electron microscope, the length of the template corresponds to the unit length of SV40 form-I DNA (263). This, therefore, rules out the first possibility. The major initiation site for late transcription is located at or very near to the 0.67 map unit (263). This seems to exclude the second alternative, which needs more than one initiation site. Under the electron microscope, nascent RNA molecules up to 500 nucleotides in length appear to have their origin at or very close to the

0.67 map unit, but the leader sequences of 16 S mRNA molecules contain about 200 nucleotides that come from the 0.67 map unit region (*310*). In addition, analyses of R-loop structures formed between nuclear viral RNA and linear SV40 DNA clearly show sequences in the nuclear RNA complementary to the IVS (*315a*). This eliminates the third alternative. Therefore, it appears more plausible that the joining of leader sequences to the coding region occurs by the looping out of the intervening RNA sequences and excision, followed by ligation.

Since the discovery of splicing as a major posttranscriptional processing of eukaryotic cellular and viral mRNAs (*316*), numerous studies have attempted to determine the sequence and conformation of various sites that may act as signals for splicing and the effect of removing the IVS from the DNA template on the mRNA product. From the sequence analysis of the boundaries in a few evolutionary distant organisms, it appears that the sequence GU/AG is highly conserved at the spliced site, i.e., at the 5' and 3' end of the introns, respectively (*316, 317*). However, it seems unlikely that this dinucleotide sequence provides the only signal for splicing, since these sequences are sometimes present inside the exons that are not spliced out (*316*).

Whatever the actual mechanism of splicing, it seems likely that the ends of the IVS must be brought close to each other prior to endonucleolytic cleavage so that the 5' and 3' ends could be easily ligated, but how it is done is not yet understood. The introns possess no marked homology or symmetry of sequences among which Watson–Crick base-pairing is possible. Proflavine, an intercalating dye, inhibits the cleavage of viral nuclear RNA precursors, but the turnover of the viral mRNAs in the cytoplasm is not inhibited (*318*). Thus, the secondary and tertiary structures of RNA may be the important processing signals in the generation of viral mRNAs (*318, 319*). As suggested recently, poly(A) may also facilitate splicing by forming a triple-stranded structure within the precursor RNA at the spliced site (*320*).

It has been speculated that one class of small RNA molecules found in eukaryotic cells could be involved in splicing (*321, 322*). These molecules exist as ribonucleoprotein particles and possess a periodic structure (*321–324*). Some of these RNAs, designated as class U-1, have been sequenced, and show extensive complementarity to the concensus sequence at the spliced sites (*321, 322*). This strongly suggests that these small RNA molecules may play a key role of bringing the terminals of the IVS together and thus correctly aligning the sequences to be spliced out.

Studies with deletion mutants lacking the late region from 0.72–

0.82 map units that contains the IVS for the late 19 S mRNA plus the flanking sequences show a strong polar effect on the distant VP1 gene because of a defect in the synthesis of stable mRNA (325). From other studies, it appears that the spliced mRNAs in the cytoplasm are derived from the colinear transcripts containing the IVS (247). These suggest that spliced junctions or the IVS regions, or both, serve as control signals for the posttranscriptional processing of stable viral mRNA. This possibility was checked with IVS deletion mutants in which the late viral gene region was replaced with a cDNA copy of the late 16 S mRNA. By both complementation analysis and immunoprecipitation of viral proteins it was shown that this mutant, which lacks precisely the 16 S IVS, is unable to encode the VP-1 function and no stable mRNA was detected. These observations indicate that splicing is required not only for removing the IVS but also for stabilizing the mRNAs (326, 327). It is still not known whether the whole IVS of SV40 mRNAs is removed at one step or in parts, though the latter is true in adenovirus 2, globin and ovalbumin nuclear RNAs (328–330). Future studies with other cellular and viral systems, isolation and purification of enzyme complex(es) involved in splicing, and reconstitution of *in vitro* systems with precursor RNAs, will elucidate the molecular mechanisms of splicing and its crucial role in the regulation of eukaryotic genes.

V. The Minichromosome—A Model for the Structure and Function of Eukaryotic Chromatin

After the discovery of the subunit structure of eukaryotic chromatin (reviewed in (180, 181), biophysical and biochemical studies revealed that the SV40 genome is organized as a minichromosome having a subunit organization and protein composition similar to that of eukaryotic chromatin (16, 18). This finding opened a new avenue for the use of the SV40 minichromosome, which can be isolated intact, as a model for the study of the structure and function of eukaryotic chromatin.

Since the SV40 genome is circular, it is possible to demonstrate that the nucleosomal DNA is under a torsional constraint equivalent to -1.0 to -1.25 superhelical turns in each nucleosome (24). This is clearly observed when histones are removed from the closed circular minichromosome that appears to be relaxed. This estimate is in apparent contradiction to that obtained from X-ray diffraction studies of crystals of nucleosome core particles, namely, about 1.75 turns of DNA superhelix per 140 base-pairs associated with the histone core

(*182*). However, the apparent discrepancy is solved if one assumes that the pitch of the DNA helix changes when free DNA in solution is wrapped around the histone core. Digestion with DNase I suggests that the pitch of the DNA duplex is very close to 10 base-pairs in the nucleosome core (*183*). An increase in the number of base-pairs per helical turn to a value of 10.4–10.7 is required for DNA in solution to reconcile the two sets of results. In fact, this is in good agreement with a theoretical estimate for linear DNA in solution (*331*).

At least three different forms of the minichromosome can be isolated (*57–61a*), none of which is identical to the relaxed 55 S nucleoprotein complex reported in many earlier studies. Different forms differ in their superstructure and protein content and in the degree of histone acetylation. The interconversion of the different forms of the minichromosome can be represented as follows:

$$95\text{ S} \longrightarrow 75\text{ S} \xrightarrow{+\text{ capsid proteins}} 180\text{ S}{-}200\text{ S} \xrightarrow{+\text{ capsid proteins}} 240\text{ S} \xrightarrow{-H1} 240\text{ S}$$

(replicating minichromosome) — (minichromosome) — — (previrion) — (mature virion)

All forms of the minichromosome except the mature virion contain histone H1. It is not known why and how histone H1, which is responsible for the higher-order packaging of cellular chromatin (*180, 181*), is eliminated at the final step.

Capsid proteins in varying proportion are associated with the minichromosome isolated either from infected cells or from mature virions (*21, 31, 35*). Recent studies suggest that these proteins have some definite roles in the regulation of the expression of early genes of SV40 (*35*; G. R. Stark, personal communication). On the other hand, the presence of capsid proteins, which probably play a role partly similar to that of nonhistone proteins of cellular chromatin, may limit the use of the minichromosome as models in *in vitro* assay systems. The heterogeneity of the minichromosome also calls for caution in its use in biochemical experiments.

Although the internal architecture of the nucleosome is understood, the higher-order organization of cellular chromatin still remains unclear, and has often been approached with SV40 chromatin as a model (*23*). However, because of the circular topology of the SV40 genome and the involvement of capsid proteins in its higher order organization, it is doubtful whether it could adequately serve as a model for understanding the supranucleosomal arrangement of cellular chromatin.

The replicating SV40 genome bears similarities to individual eukaryotic replicons and contains nucleosomes on both sides of the "eye" (*192, 193*). Replication is bidirectional in both cases and proceeds from a unique origin in the former, but it is not known whether the origin in mammalian DNA is unique for individual replicons. Many aspects of SV40 replication, including the semidiscontinuous mode of DNA synthesis, RNA priming, basic enzymology, and mechanism of synthesis of Okazaki pieces and their maturation, segregation of histones, etc., are very similar to those of cellular systems (*152, 158*). However, the exact mechanism of initiation of DNA replication is not understood in any of these systems.

In SV40, a viral gene product, T antigen, which binds preferentially with a sequence of about 120 base-pairs containing the origin of replication, is essential in the initiation of the replication cycle, but it is not clear whether it is the DNA-binding property of the T antigen or an associated enzymic activity that is responsible for the initiation of replication.

Using a few basic criteria, it is possible to show that, like mammalian systems, the SV40 genome in isolated nuclei can be made to continue DNA synthesis *in vitro* (*91*). It involves the synthesis and joining of Okazaki fragments identical to that *in vivo*, and normal bidirectional replication until the two replication forks merge in the termination region. The only feature not reproduced *in vitro* is the initiation of replication.

The rate of conversion of the SV40 replicating genome into a mature one is about 10 to 20 times slower than expected, indicating the accumulation of replicating intermediates at some points late during replication. In a recent model of cellular DNA replication, it is suggested that the growing points of the replicon are attached, presumably through DNA or nonhistone chromatin protein to nuclear matrix, and the DNA is reeled through these sites (*92*). To reconcile a fixed site of replication with bidirectional replication, there must be two adjacent fixed replication complexes for each, which may be equivalent to the topological constraint of closed circular DNA. It is still not understood whether the segregation of the daughter strands of the cellular and SV40 DNA employs the same mechanism.

Only about 10% of the eukaryotic genome is functionally active, and this active chromatin shares many properties with its inactive counterpart (*332*). Although earlier electron microscopic studies did not show any clear-cut evidence for the presence of nucleosomes on actively transcribing ribosomal genes, recent studies indicate the presence of nucleosome-like beads spaced irregularly at large intervals (333–335). Whether the beads contain the full complement of histones and their degree of acetylation are not known. The active chromatin yields a digestion pattern with microccocal nuclease or DNase II that is typical of a subunit structure but differs distinctly in its hypersensitivity to DNase I, higher degree of histone acetylation, and in the presence of smaller amounts of histone H1 (332). The basis of altered nucleosomal conformation, as detected by hypersensitivity to DNase I, is also not understood. It is thought that HMG 14 and 17 together with other factors may play a role in making the active chromatin extremely sensitive to DNase I (336, 337).

In contrast to cellular chromatin, about all of the SV40 genome is functionally active and contains only five genes. The active SV40 transcriptional complex isolated late in infection contains a minichromosomal template (286), and it provides the only evidence, to our knowledge, that non-rDNA contains nucleosomes during its active transcription. No HMG proteins are detected, either in mature virions or in the minichromosome isolated from infected cell nuclei. The degree of histone acetylation in the minichromosome is higher than that in the host cell, and it increases progressively with the maturation of the minichromosome (36, 39). Not all of the SV40 DNA is organized into nucleosomes; about 19% of the DNA is present as nonnucleosomal DNA (42). The significance of this nonnucleosomal DNA, which is also present in active cellular chromatin, is not clear at present.

The molecular mechanisms controlling the specificity of transcription in eukaryotic systems is still far from understood. One of the most sensitive assays for the specificity of transcription involves transcription of specific sequences. There are many reports on the transcription of ovalbumin genes from oviduct chromatin (338), but most of these studies suffer from problems caused by the endogenous RNA contaminant in the chromatin, and the difficulty of measuring the contribution of this endogenous RNA to the total RNA synthesized. Attempts to separate the *in vitro* transcript by labeling with Hg-[^{3}H]UTP have been only partially successful (338). To overcome these problems, the SV40 minichromosome is being used extensively as a model to understand how the eukaryotic genome is transcribed.

The use of the SV40 minichromosome in the *in vitro* assay system is appealing for at least two reasons: (*a*) the minichromosome is transcribed by the host RNA polymerase II that transcribes cellular genes (*12*); (*b*) SV40 mRNAs are processed *in vivo* like other cellular mRNAs (*13, 14*). As discussed earlier, viral transcriptional complexes isolated from infected cells or RNA-free minichromosomes from mature virions can be used as templates. In such systems, the nucleosomal

DNA is available for transcription (284–292) and the overall conformation of the nucleosome remains unaltered after transcription. The efficiency of transcription is increased by conditions such as increased ionic strength that loosen DNA-histone interactions, but the specificity of transcription of the SV40 genome *in vitro* has not been achieved, probably because of the absence of specific factor(s) either in the template or in the polymerase. We hope that future studies with SV40 minichromosomes will enable reconstitution of a system for specific transcription and promote understanding of other basic aspects of eukaryotic gene replication. (See Note Added in Proof.)

ACKNOWLEDGMENTS

Thanks are expressed to various investigators for sending us their preprints and unpublished observations. The authors are indebted to Drs. E. Bernstine, R. K. Fujimura, and S. Mitra for critical review of the manuscript and to Mrs. Rose Feldman for her help in proofreading. We are especially grateful to Dr. Waldo E. Cohn for his critical suggestions and incisive comments.

REFERENCES

1. J. T. Finch and A. Klug, *JMB* **13**, 1 (1965).
2. F. A. Anderer, H. D. Schlumberger, M. A. Koch, H. Frank and H. J. Eggers, *Virology* **32**, 511 (1967).
3. V. B. Reddy, B. Thimmappaya, R. Dhar, K. N. Subramanian, B. S. Zain, J. Pan, P. K. Ghosh, M. L. Celma and S. M. Weissman, *Science* **200**, 494 (1978).
4. W. Fiers, R. Contreras, G. Haegeman, R. Rogiers, A. Van de Voorde, H. Van Heuverswyn, J. Van Herreweghe, G. Volckaert and M. Ysebaert, *Nature* **273**, 113 (1978).
5. T. J. Kelly, Jr. and D. Nathans, *Adv. Virus Res.* **21**, 85 (1977).
6. J. Sambrook, *in* "The Molecular Biology of Animal Viruses" (D. P. Nayak, ed.), Vol. 2, p. 589. Dekker, New York, 1978.
6a. D. Nathans, *Science* **206**, 903 (1979).
7. A. F. Purchio and G. C. Fareed, *in* "Methods in Enzymology," Vol. 68, p. 357. Academic Press, New York, 1979.
8. N. P. Salzman and G. Khoury, *in* "Comprehensive Virology" (H. Fraenkel-Conrat and R. Wagner, eds.), Vol. 3, p. 63. Plenum, New York, 1974.
9. G. C. Fareed and D. Davoli, *ARB* **46**, 471 (1977).
10. P. Lebowitz and S. M. Weissman, *Curr. Top. Microbiol. Immunol.* **87**, 43 (1979).
11. S. Mitra, *Annu. Rev. Genet.* **14**, 347 (1980).
12. A. H. Jackson and B. Sugden, *J. Virol.* **10**, 1086 (1972).
13. M. Revel and Y. Groner, *ARB* **47**, 1079 (1978).
14. J. Abelson, *ARB* **48**, 1035 (1979).
15. F. Crick, *Science* **204**, 264 (1979).
16. J. D. Griffith, *Science* **187**, 1202 (1975).
17. A. L. Olins and D. E. Olins, *Science* **183**, 330 (1974).
18. C. Crémisi, P. F. Pignatti, O. Croissant and M. Yaniv, *J. Virol.* **17**, 204 (1976).
19. G. C. Das, D. P. Allison and S. K. Niyogi, *BBRC* **89**, 17 (1979).
20. G. Christiansen and J. Griffith, *NARes* **4**, 1837 (1977).
21. U. Müller, H. Zentgraf, I. Eicken and W. Keller, *Science* **201**, 406 (1978).

22. A. J. Varshavsky, S. A. Nedospasov, V. V. Schmatchenko, V. V. Bakayev, P. M. Chumackov and G. P. Georgiev, *NARes* **4**, 3303 (1977).
23. A. J. Varshavsky, V. V. Bakayev, S. A. Nedospasov and G. P. Georgiev, *CSHSQB* **42**, 457 (1978).
24. J. E. Germond, B. Hirt, P. Oudet, M. Gross-Bellard and P. Chambon, *PNAS* **72**, 1843 (1975).
25. W. Meinke, M. R. Hall and D. A. Goldstein, *J. Virol.* **15**, 439 (1975).
26. K. B. Tan and F. Sokol, *J. Virol.* **10**, 985 (1972).
27. D. M. Pett, M. K. Estes and J. S. Pagano, *J. Virol.* **15**, 379 (1975).
28. S. A. Nedospasov, V. V. Bakayev and G. P. Georgiev, *NARes* **5**, 2847 (1978).
29. A. J. Varshavsky, V. V. Bakayev, P. M. Chumackov and G. P. Georgiev, *NARes* **3**, 2101 (1976).
30. J. P. MacGregor, Y. H. Chen, D. A. Goldstein and M. R. Hall, *BBRC* **83**, 814 (1978).
31. M. Coca-Prados and M.-T. Hsu, *J. Virol.* **31**, 199 (1979).
32. K. B. Tan, *PNAS* **74**, 2805 (1977).
33. Y.-H. Chen, J. P. MacGregor, D. A. Goldstein and M. R. Hall, *J. Virol.* **30**, 218 (1979).
34. V. V. Schmatchenko and A. J. Varshavsky, *Anal. Biochem.* **85**, 42 (1978).
35. J. N. Brady, C. Lavialle and N. P. Salzman, *J. Virol.* **35**, 371 (1980).
36. F. La Bella, G. Vidali and C. Vesco, *Virology* **96**, 564 (1979).
37. A. Chestier and M. Yaniv, *PNAS* **76**, 46 (1979).
38. B. S. Schaffhausen and T. L. Benjamin, *PNAS* **73**, 1092 (1976).
39. F. La Bella and C. Vesco, *J. Virol.* **33**, 1138 (1980).
40. M. Bellard, P. Oudet, J. E. Germond and P. Chambon, *EJB* **70**, 543 (1976).
41. B. A. J. Ponder, F. Crew and L. V. Crawford, *J. Virol.* **25**, 175 (1978).
41a. M. Noll, *Nature* **251**, 249 (1974).
42. E. R. Shelton, P. M. Wassarman and M. L. DePamphilis, *JMB* **125**, 491 (1978).
42a. H. A. Feldman, E. R. Shelton, P. M. Wassarman, and M. L. DePamphilis, *JMB* **125**, 511 (1978).
43. B. Polisky and B. McCarthy, *PNAS* **72**, 2895 (1975).
44. C. Crémisi, P. F. Pignatti and M. Yaniv, *BBRC* **73**, 548 (1976).
45. B. A. J. Ponder and L. V. Crawford, *Cell* **11**, 35 (1977).
46. M. Persico-DiLauro, R. G. Martin and D. M. Livingston, *J. Virol.* **24**, 451 (1977).
47. M. L. DePamphilis, R. Hay, E. Shelton, L. Tack, D. Tapper, D. Weaver, and P. M. Wassarman, *FP* **39**, 2121 (1980).
48. A. J. Varshavsky, O. H. Sundin and M. J. Bohn, *NARes* **5**, 3469 (1978).
49. A. J. Varshavsky, O. H. Sundin and M. J. Bohn, *Cell* **16**, 453 (1979).
50. W. A. Scott and D. J. Wigmore, *Cell* **15**, 1511 (1978).
51. W. Waldeck, B. Föhring, K. Chowdhury, P. Gruss and G. Sauer, *PNAS* **75**, 5964 (1978).
52. S. Saragosti, G. Moyne and M. Yaniv, *Cell* **20**, 65 (1980).
52a. E. B. Jakobovits, S. Bratosin and Y. Aloni, *Nature* **285**, 263 (1980).
53. O. Sundin and A. Varshavsky, *JMB* **132**, 535 (1979).
54. S. A. Nedospasov and G. P. Georgiev, *BBRC* **92**, 532 (1980).
55. J. Griffith, M. Dieckmann and P. Berg, *J. Virol.* **15**, 167 (1975).
56. H. Kasamatsu and M. Wu, *BBRC* **68**, 927 (1976).
57. I. Baumgartner, C. Kuhn and E. Fanning, *Virology* **96**, 54 (1979).
58. R. Fernandez-Munoz, M. Coca-Prados and M.-T. Hsu, *J. Virol.* **29**, 612 (1979).
59. E. A. Garber, M. M. Seidman and A. J. Levine, *Virology* **90**, 305 (1978).
60. K. H. Klempnauer, E. Fanning, B. Otto and R. Knipers, *JMB* (in press).

61. E. Fanning and I. Baumgartner, *Virology* **102**, 1 (1980).
61a. E. B. Jakobovits and Y. Aloni, *Virology* **102**, 107 (1980).
62. M. M. Seidman, E. A. Garber and A. J. Levine, *Virology* **95**, 256 (1979).
63. H. L. Ozer and P. Tegtmeyer, *J. Virol.* **9**, 52 (1972).
64. J. N. Brady, V. D. Winston and R. A. Consigli, *J. Virol.* **27**, 193 (1978).
65. G. C. Das and S. K. Niyogi, *BBRC* **89**, 1267 (1979).
66. J. N. Brady, J. D. Kendall and R. A. Consigli, *J. Virol.* **32**, 640 (1979).
67. A. Worcel, *CSHSQB* **42**, 313 (1978).
68. R. G. Martin, *Virology* **83**, 433 (1977).
69. G. C. Fareed, C. F. Garon and N. P. Salzman, *J. Virol.* **10**, 484 (1972).
70. K. J. Danna and D. Nathans, *PNAS* **69**, 3097 (1972).
71. K. N. Subramanian and T. Shenk, *NARes* **5**, 3635 (1978).
72. M. W. Gutai and D. Nathans, *JMB* **126**, 259 (1978).
73. K. N. Subramanian, R. Dhar and S. M. Weissman, *JBC* **252**, 355 (1977).
74. D. Shortle and D. Nathans, *PNAS* **75**, 2170 (1978).
75. D. Shortle and D. Nathans, *JMB* **131**, 801 (1979).
76. P. Tegtmeyer, *J. Virol.* **10**, 591 (1972).
77. K. Mann and T. Hunter, *J. Virol.* **29**, 232 (1979).
78. A. J. Levine, P. C. van der Vliet and J. S. Sussenbach, *Curr. Top. Microbiol. Immunol.* **73**, 67 (1979).
79. C.-J. Lai and D. Nathans, *JMB* **97**, 113 (1975).
80. B. E. Griffin, M. Fried and A. Cowie, *PNAS* **71**, 2077 (1974).
81. M. A. Martin, L. D. Gelb, C. Garon, K. K. Takemoto, T. N. H. Lee, G. H. Sack, Jr. and D. Nathans, *Virology* **59**, 179 (1974).
82. M. A. Martin, G. Khoury and G. C. Fareed, *CSHSQB* **39**, 129 (1975).
83. W. W. Brockman, T. N. H. Lee and D. Nathans, *CSHSQB* **39**, 119 (1975).
84. W. W. Brockman, M. W. Gutai and D. Nathans, *Virology* **66**, 36 (1975).
85. G. C. Fareed, M. L. McKerlie and N. P. Salzman, *JMB* **74**, 95 (1973).
86. P. J. Laipis, A. Sen, A. J. Levine and C. Mulder, *Virology* **68**, 115 (1975).
87. M. C. Y. Chen, E. Birkenmeier and N. P. Salzman, *J. Virol.* **17**, 614 (1976).
88. R. Jaenisch and A. J. Levine, *JMB* **73**, 199 (1973).
89. M. M. Seidman and N. P. Salzman, *J. Virol.* **30**, 600 (1979).
90. D. P. Tapper and M. L. DePamphilis, *JMB* **120**, 401 (1978).
91. D. P. Tapper, S. Anderson and M. L. DePamphilis, *BBA* **565**, 84 (1979).
92. D. M. Pardoll, B. Vogelstein and D. S. Coffey, *Cell* **19**, 527 (1980).
93. R. Okazaki, T. Okazaki, K. Sakabe, K. Sugimoto and A. Sugino, *PNAS* **59**, 598 (1968).
94. B. Alberts and R. Sternglanz, *Nature* **269**, 655 (1977).
95. G. C. Fareed and N. P. Salzman, *Nature NB* **238**, 274 (1972).
96. J. A. Huberman and H. Horwitz, *CSHSQB* **38**, 233 (1974).
97. G. C. Fareed, G. Khoury and N. P. Salzman, *JMB* **77**, 457 (1973).
98. G. Magnusson, *J. Virol.* **12**, 600 (1973).
99. V. Pigiet, R. Eliasson and P. Reichard, *JMB* **84**, 197 (1974).
100. B. Francke and T. Hunter, *JMB* **83**, 99 (1974).
101. B. Francke and M. Vogt, *Cell* **5**, 205 (1975).
102. D. Perlman and J. A. Huberman, *Cell* **12**, 1029 (1978).
103. D. L. Robberson, L. V. Crawford, C. Syrett and A. W. James, *J. Gen. Virol.* **26**, 59 (1975).
104. M. Schnös and R. B. Inman, *JMB* **51**, 61 (1970).
105. I. R. Lehman, B.-K. Tye and P.-O. Nyman, *CSHSQB* **43**, 221 (1978).

106. K. Brynolf, R. Eliasson and P. Reichard, *Cell* **13**, 573 (1978).
107. T. Hunter, B. Francke and L. Bacheler, *Cell* **12**, 1021 (1977).
108. P. J. Flory, Jr., *NARes* **4**, 1449 (1977).
109. P. Reichard, R. Eliasson and G. Söderman, *PNAS* **71**, 4901 (1974).
110. R. Eliasson and P. Reichard, *JBC* **253**, 7469 (1978).
111. S. Anderson, G. Kaufmann and M. L. DePamphilis, *Bchem* **16**, 4990 (1977).
112. G. Kaufmann, S. Anderson and M. L. DePamphilis, *JMB* **116**, 549 (1977).
113. N. P. Salzman and M. M. Thoren, *J. Virol.* **11**, 721 (1973).
114. H. Krokan, P. Schaffer and M. L. DePamphilis, *Bchem* **18**, 4431 (1979).
115. H. J. Edenberg, S. Anderson and M. L. DePamphilis, *JBC* **253**, 3273 (1978).
116. S. Hirose, R. Okazaki and F. Tamanoi, *JMB* **77**, 501 (1973).
117. T. M. Herman, M. L. DePamphilis and P. M. Wassarman, *Bchem* **18**, 4563 (1979).
118. C. Prives, E. Gilboa, M. Revel and E. Winocour, *PNAS* **74**, 457 (1977).
119. L. V. Crawford, C. N. Cole, A. E. Smith, E. Paucha, P. Tegtmeyer, K. Rundell and P. Berg, *PNAS* **75**, 117 (1978).
120. E. May, M. Kress and P. May, *NARes* **5**, 3083 (1978).
121. M. J. Sleigh, W. C. Topp, R. Hanich and J. F. Sambrook, *Cell* **14**, 79 (1978).
122. R. Seif and R. G. Martin, *J. Virol.* **32**, 979 (1979).
123. N. Bouck, N. Beales, T. Shenk, P. Berg and G. diMayorca, *PNAS* **75**, 2473 (1978).
124. P. Tegtmeyer, *J. Virol.* **15**, 613 (1975).
125. M. Graessmann and A. Graessmann, *PNAS* **73**, 366 (1976).
126. K. Cowan, P. Tegtmeyer and D. D. Anthony, *PNAS* **70**, 1927 (1973).
127. S. I. Reed, G. R. Stark and J. C. Alwine, *PNAS* **73**, 3083 (1976).
128. G. Kimura and R. Dulbecco, *Virology* **52**, 529 (1973).
129. J. S. Brugge and J. S. Butel, *J. Virol.* **15**, 619 (1975).
130. R. J. Martin and J. Y. Chou, *J. Virol.* **15**, 599 (1975).
131. M. Osborn and K. Weber, *J. Virol.* **15**, 636 (1975).
132. R. Tjian, G. Fey and A. Graessmann, *PNAS* **75**, 1279 (1978).
133. R. B. Carroll, L. Hager and R. Dulbecco, *PNAS* **71**, 3754 (1974).
134. S. I. Reed, J. Ferguson, R. W. Davis and G. R. Stark, *PNAS* **72**, 1605 (1975).
135. D. Jessel, T. Landau, J. Hudson, T. Lalor, D. Tenen and D. M. Livingston, *Cell* **8**, 535 (1976).
136. T. Spillman, D. Giacherio and L. P. Hager, *JBC* **254**, 3100 (1979).
137. R. Tjian, *CSHSQB* **43**, 655 (1979).
138. M. Oren, E. Winocour and C. Prives, *PNAS* **77**, 220 (1980).
139. R. Tjian and A. Robbins, *PNAS* **76**, 610 (1979).
140. D. R. Shortle, R. F. Margolskee and D. Nathans, *PNAS* **76**, 6128 (1979).
141. R. Tjian and A. Robbins, *CSHSQB* Viral Oncogens, May 30–June 6 (1979).
142. J. D. Griffin, G. Spangler and D. M. Livingston, *PNAS* **76**, 2610 (1979).
143. D. Giacherio and L. P. Hager, *JBC* 254, 8113 (1979).
144. J. Reiser, J. Renart, L. V. Crawford and G. R. Stark, *J. Virol.* **33**, 78 (1980).
145. M. Méchali, M. Girard and A.-M. De Recondo, *J. Virol.* **23**, 117 (1973).
146. R. W. Chiu and E. F. Baril, *JBC* **250**, 7951 (1975).
147. S. Riva, M. A. Cline and P. J. Laipis, *in* "DNA Synthesis. Present and Future" (I. Molineux and M. Kohiyama, eds.), pp. 641–654. Plenum, New York, 1977.
148. S. Spadari, G. Villani and N. Hardt, *Exp. Cell Res.* **113**, 57 (1978).
149. B. Otto, E. Fanning and A. Richter, *CSHSQB* **43**, 705 (1979).
150. Y. Tsubota, M. A. Waqar, J. F. Burke, B. I. Milavetz, M. J. Evans, D. Kowalski and J. A. Huberman, *CSHSQB* **43**, 693 (1979).
151. B. Otto and E. Fanning, *NARes* **5**, 1715 (1978).

152. M. L. DePamphilis, S. Anderson, R. Bar-Shavit, E. Collins, H. Edenberg, T. Herman, B. Karas, G. Kaufmann, H. Krokan, E. Shelton, R. Su, D. Tapper and P. M. Wassarman, *CSHSQB* **43**, 679 (1979).
153. A. Kornberg, "DNA Replication." Freeman, San Francisco, California, 1980.
154. S. Yoshida, M. Yamada and S. Masaki, *BBA* **477**, 144 (1977).
155. G. Magnusson, V. Pigiet, E. L. Winnacker, R. Abrams and P. Reichard, *PNAS* **70**, 412 (1973).
155a. P. J. Laipis and A. J. Levine, *Virology* **56**, 580 (1973).
156. G. Magnusson, R. Craig, M. Närkhammar, P. Reichard, M. Staub and H. Warner, *CSHSQB* **39**, 227 (1975).
157. M. A. Waqar, M. J. Evans and J. A. Huberman, *NARes* **5**, 1933 (1978).
158. T. Okazaki, Y. Kurosawa, T. Ogawa, T. Seki, K. Shinozaki, S. Hirose, A. Fujiyama, Y. Kohara, Y. Machida, F. Tamanoi and T. Hozumi, *CSHSQB* **43**, 203 (1979).
159. T. Hunter and B. Francke, *JMB* **83**, 123 (1974).
160. P. Chambon, *ARB* **44**, 613 (1975).
161. R. Eliasson and P. Reichard, *Nature* **272**, 184 (1978).
162. J. G. Stavrianopoulos, A. Gambino-Giuffrida and E. Chargaff, *PNAS* **73**, 1087 (1976).
163. T. Lindahl and G. M. Edelman, *PNAS* **61**, 680 (1968).
164. P. Beard, *BBA* **269**, 385, (1972).
165. G. C. F. Pederaliony, S. Spadar, G. Ciarrocchi, A. M. Pedrini and A. Falaschi, *EJB* **39**, 343 (1973).
166. S. Soderhall and T. Lindahl, *JBC* **248**, 672 (1973).
167. N. R. Cozzarelli, *Science* **207**, 953 (1980).
168. J. C. Wang and L. F. Liu, *in* "Molecular Genetics" (J. H. Taylor, ed.), Part 3, p. 65. Academic Press, New York, 1979.
169. C. Hamelin and M. Yaniv, *NARes* **7**, 679 (1979).
170. J. E. Germond, J. Rouvière-Yaniv, M. Yaniv and D. Brutlag, *PNAS* **76**, 3779 (1979).
171. L. F. Liu, C.-C. Liu and B. M. Alberts, *Cell* **19**, 697 (1980).
172. B. M. Alberts and L. Frey, *Nature* **227**, 1313 (1970).
173. G. Herrick, H. Delius and B. Alberts, *JBC* **251**, 2142 (1976).
174. K. B. Tan, *PNAS* **74**, 2805 (1977).
175. S. C. R. Elgin and H. Weintraub, *ARB* **44**, 725 (1975).
176. R. Hancock, *JMB* **40**, 457 (1969).
177. R. D. Kornberg, *Science* **184**, 868 (1974).
178. P. Oudet, M. Gross-Bellard and P. Chambon, *Cell* **4**, 281 (1975).
179. P. Chambon, *CSHSQB* **42**, 1209 (1978).
180. G. Felsenfeld, *Nature* **271**, 115 (1978).
181. R. D. Kornberg, *ARB* **46**, 931 (1977).
182. J. T. Finch, L. C. Lutter, D. Rhodes, R. S. Brown, B. Rushton, M. Levitt and A. Klug, *Nature* **269**, 29 (1977).
183. L. C. Lutter, *CSHSQB* **42**, 137 (1978).
184. I. M. Leffak, R. Grainger, and H. Weintraub, *Cell* **12**, 837 (1977).
185. H. Weintraub, A. Worcel and B. Alberts, *Cell* **9**, 409 (1976).
186. B. Richards, J. Pardon, D. Lilley, R. Cotter and J. Wooley, *Cell Biol. Int. Rep.* **1**, 107 (1977).
187. P. Oudet, J.-E. Germond, M. Bellard, C. Spadafora and P. Chambon, *Philos. Trans. R. Soc. London B* **283**, 241 (1978).
188. C. Crémisi and M. Yaniv, *BBRC* **92**, 1117 (1980).
189. A. Worcel, S. Han and M. L. Wong, *Cell* **15**, 969 (1978).
190. F. X. Wilhelm, M. L. Wilhelm, M. Erard and M. Daune, *NARes* **5**, 505 (1978).

191. C. Crémisi, *Microbiol. Rev.* **43**, 297 (1979).
192. S. L. McKnight and O. L. Miller, Jr., *Cell* **12**, 795 (1977).
193. C. Crémisi, A. Chestier and M. Yaniv, *CSHSQB* **42**, 409 (1978).
194. M. M. Seidman, C. F. Garon, and N. P. Salzman, *NARes* **5**, 2877 (1978).
195. V. Jackson, D. K. Granner and P. Chalkey, *Bchem* **73**, 2266 (1976).
196. R. Hancock, *PNAS* **75**, 2130 (1978).
197. R. L. Seale, *PNAS* **73**, 2270 (1976).
198. H. Weintraub, *Cell* **9**, 419 (1976).
199. D. Riley and H. Weintraub, *PNAS* **76**, 328 (1979).
200. R. Seale, *Cell* **9**, 423 (1976).
201. M. M. Seidman, A. J. Levine and H. Weintraub, *Cell* **18**, 439 (1979).
202. C. E. Hildebrand and R. A. Walters, *BBRC* **73**, 157 (1976).
203. J. E. Schlaeger and K.-H. Klempnauer, *EJB* **89**, 567 (1978).
204. R. A. Laskey, B. M. Honda, A. D. Mills, and J. T. Finch, *Nature* **275**, 416 (1978).
205. T. Nelson, T.-S. Hsieh and D. Brutlag, *PNAS* **76**, 5510 (1979).
206. G. C. Das, P. C. McGray, D. P. Allison and S. K. Niyogi, *FP* **39**, 2196 (1980).
207. A. Stein, J. P. Whitlock, Jr. and M. Bina, *PNAS* **76**, 5000 (1979).
208. H. Westphal, *JMB* **50**, 407 (1970).
209. H. Westphal and E. D. Kiehn, *CSHSQB* **35**, 819 (1971).
210. D. M. Lindstrom and R. Dulbecco, *PNAS* **69**, 1517 (1972).
211. G. Khoury, J. C. Byrne and M. A. Martin, *PNAS* **69**, 1925 (1972).
212. G. Khoury and M. A. Martin, *Nature NB* **238**, 4 (1972).
213. J. Sambrook, P. A. Sharp and W. Keller, *JMB* **70**, 57 (1972).
214. J. Sambrook, B. Sugden, W. Keller and P. A. Sharp, *PNAS* **70**, 3711 (1973).
215. B. S. Zain, R. Dhar, S. M. Weissman, P. Lebowitz and A. M. Lewis, Jr., *J. Virol.* **11**, 682 (1973).
216. R. Dhar, K. Subramanian, B. S. Zain, J. Pan and S. Weissman, *CSHSQB* **39**, 153 (1975).
217. P. Lebowitz, R. Stern, P. K. Ghosh and S. M. Weissman, *J. Virol.* **22**, 430 (1977).
218. T. L. Benjamin, *JMB* **16**, 359 (1966).
219. Y. Aloni, E. Winocour, and L. Sachs, *JMB* **31**, 415 (1968).
220. K. Oda and R. Dulbecco, *PNAS* **60**, 525 (1968).
221. S. Tonegawa, G. Walter, A. Bernardini and R. Dulbecco, *CSHSQB* **35**, 823 (1971).
222. R. Weil, C. Salomon, E. May and P. May. *In* "Viruses, Evolution and Cancer" (E. Kurstak and K. Maramorosch, eds.), p. 455. Academic Press, New York, 1974.
223. S. I. Reed and J. C. Alwine, *Cell* **11**, 523 (1977).
224. G. Khoury, M. A. Martin, T. N. H. Lee, K. J. Danna and D. Nathans, *JMB* **78**, 377 (1973).
225. G. Khoury, P. Howley, D. Nathans and M. Martin, *J. Virol.* **15**, 433 (1975).
226. N. H. Acheson, *Cell* **8**, 1 (1976).
227. P. Beard, N. H. Acheson and I. H. Maxwell, *J. Virol.* **17**, 20 (1976).
228. N. P. Salzman, N. Birkenmeier, N. Chiu, E. May, M. F. Radonovich and M. Shani, *INSERM* **69**, 243 (1977).
229. F.-J. Ferdinand, M. Brown and G. Khoury, *PNAS* **74**, 5443 (1977).
230. E. H. Birkenmeier, N. Chiu, M. F. Radonovich, E. May and N. P. Salzman, *J. Virol.* **29**, 983 (1979).
231. N. H. Acheson and F. Miéville, *J. Virol.* **28**, 885 (1978).
232. B. A. Parker and G. R. Stark, *J. Virol.* **31**, 360 (1979).
233. R. A. Weinberg, Z. Ben-Ishai and J. E. Newbold, *Nature NB* **238**, 111 (1972).
234. R. Dhar, K. N. Subramanian, B. S. Zain, A. Levine, C. Patch and S. M. Weissmann, *INSERM* **47**, 25 (1975).

235. R. Dhar, K. N. Subramanian, J. Pan and S. M. Weissman, *JBC* **252**, 368 (1977).
236. A. J. Berk and P. A. Sharp, *PNAS* **75**, 1274 (1978).
237. E. Paucha, R. Harvey and A. E. Smith, *J. Virol.* **28**, 154 (1978).
238. N. H. Acheson, E. Buetti, K. Scherrer and R. Weil, *PNAS* **68**, 2231 (1971).
239. Y. Aloni, *CSHSQB* **39**, 165 (1974).
240. Y. Aloni, M. Shani and Y. Reuveni, *PNAS* **72**, 2587 (1975).
241. O. Laub and Y. Aloni, *J. Virol.* **16**, 1171 (1975).
242. S. Rozenblatt and E. Winocour, *Virology* **50**, 558 (1972).
243. M. A. Martin and J. C. Byrne, *J. Virol.* **6**, 463 (1970).
244. R. A. Weinberg, S. O. Warnaar and E. Winocour, *J. Virol.* **10**, 193 (1972).
245. Y. Aloni, *PNAS* **69**, 2404 (1972).
246. A. H. Fried, *Virology* **88**, 286 (1978).
247. C.-J. Lai, R. Dhar, and G. Khoury, *Cell* **14**, 971 (1978).
248. J. P. Ford and M.-T. Hsu, *J. Virol.* **28**, 795 (1978).
249. E. Buetti, *J. Virol.* **14**, 249 (1974).
250. V. B. Reddy, R. Dhar and S. M. Weissman, *JBC* **253**, 621 (1978).
251. Y. Aloni, S. Bratosin, R. Dhar, O. Laub, M. Horowitz and G. Khoury, *CSHSQB* **42**, 559 (1978).
252. M.-T. Hsu and J. Ford, *CSHSQB* **42**, 571 (1978).
253. S. Lavi and Y. Groner, *PNAS* **74**, 5323 (1977).
254. M. L. Celma, R. Dhar, J. Pan and S. M. Weissman, *NARes* **4**, 2549 (1977).
255. S. Lavi and A. J. Shatkin, *PNAS* **72**, 2012 (1975).
256. Y. Aloni, *FEBS Lett.* **54**, 363 (1975).
257. Y. Groner, P. Carmi and Y. Aloni, *NARes* **4**, 3959 (1977).
258. G. Haegeman and W. Fiers, *J. Virol.* **25**, 824 (1978).
259. D. Canaani, C. Kahana, A. Mukamel and Y. Groner, *PNAS* **76**, 3078 (1979).
260. C. Mueller, A. Graessmann and M. Graessman, *Cell* **15**, 579 (1978).
261. C. N. Cole, T. Landers, S. P. Goff, S. Manteuil-Brutlag and P. Berg, *J. Virol.* **24**, 277 (1977).
262. T. E. Shenk, J. Carbon and P. Berg, *J. Virol.* **18**, 664 (1976).
263. O. Laub, S. Bratosin, M. Horowitz and Y. Aloni, *Virology* **92**, 310 (1979).
264. M. Chamberlin, *in* "RNA Polymerase" (R. Losick and M. Chamberlin, eds.), p. 159. Cold Spring Harbor Lab., Cold Spring Harbor, New York, 1976.
265. W. Gilbert, *ibid.*, p. 193.
266. O. Laub and Y. Aloni, *Virology* **75**, 346 (1976).
267. D. H. Metz, M. N. Oxman and M. J. Levin, *BBRC* **75**, 172 (1976).
268. J.-L. Mandel and P. Chambon, *EJB* **41**, 379 (1974).
269. J.-L. Mandel and P. Chambon, *EJB* **41**, 367 (1974).
270. P. A. Weil, D. S. Luse, J. Segall and R. G. Roeder, *Cell* **18**, 469 (1979).
271. B. Waslyk, C. Kédinger, J. Corden, O. Brison and P. Chambon, *Nature* **285**, 367 (1980).
272. D. W. Chandler and J. Gralla, *Bchem* **19**, 1604 (1980).
273. M. K. Lewis and R. R. Burgess, *JBC* **255**, 4928 (1980).
274. J. A. Thompson, J. N. Brady, M. F. Radonovich, C. Lavialle and N. P. Salzman, *FP* **39**, 2203 (1980).
275. M. Girard, L. Marty and S. Manteuil, *PNAS* **71**, 1267 (1974).
276. A. Graessmann, M. Graessmann and C. Mueller, *PNAS* **74**, 4831 (1977).
277. E. H. Birkenmeier, E. May and N. P. Salzman, *J. Virol.* **22**, 702 (1977).
278. E. H. Birkenmeier, M. F. Radonovich, M. Shani and N. P. Salzman, *Cell* **11**, 495 (1977).
279. M. H. Green and T. L. Brooks, *J. Virol.* **26**, 325 (1978).

280. J. C. Alwine and G. Khoury, *J. Virol.* **33**, 920 (1980).
281. P. Gariglio and S. Mousset, *FEBS Lett.* **56**, 149 (1975).
282. M. H. Green and T. L. Brooks, *Virology* **72**, 110 (1976).
283. F.-J. Ferdinand, M. Brown, and G. Khoury, *Virology* **78**, 150 (1977).
284. T. L. Brooks and M. H. Green, *NARes* **4**, 4261 (1977).
285. M. H. Green and T. L. Brooks, *NARes* **4**, 4279 (1977).
286. P. Gariglio, R. Llopis, P. Oudet and P. Chambon, *JMB* **131**, 75 (1979).
287. R. C. Condit, A. Cowie, R. Kamen and F. Birg, *JMB* **115**, 215 (1977).
288. C. Crémisi, A. Chestier, C. Dauguet and M. Yaniv, *BBRC* **78**, 74 (1977).
289. M. R. Hall, *BBRC* **76**, 698 (1977).
290. G. Meneguzzi, N. Chenciner and G. Milanesi, *NARes* **6**, 2947 (1979).
291. B. Wasylyk and P. Chambon, *EJB* **98**, 317 (1979).
292. B. Wasylyk, G. Thevenin, P. Oudet and P. Chambon, *JMB* **128**, 411 (1979).
293. J. C. Alwine, S. I. Reed, J. Ferguson and G. R. Stark, *Cell* **6**, 529 (1975).
294. P. Tegtmeyer, M. Schwartz, J. K. Collins and K. Rundell, *J. Virol.* **16**, 168 (1975).
295. J. A. Robb, P. Tegtmeyer, A. Ishikawa and H. L. Ozer, *J. Virol.* **13**, 662 (1974).
296. J. C. Alwine, S. I. Reed and G. R. Stark, *J. Virol.* **24**, 22 (1977).
297. G. Khoury and E. May, *J. Virol.* **23**, 167 (1977).
298. J. Avila, R. Saral, R. G. Martin and G. Khoury, *Virology* **73**, 89 (1976).
299. J. Y. Chou and R. G. Martin, *J. Virol.* **15**, 145 (1975).
300. J. A. Robb and R. G. Martin, *J. Virol.* **9**, 956 (1972).
301. W. Gilbert, *Nature* **271**, 501 (1978).
302. S. Tonegawa, A. M. Maxam, R. Tijard, O. Bernard and W. Gilbert, *PNAS* **75**, 1485 (1978).
303. V. B. Reddy, P. K. Ghosh, P. Lebowitz, M. Piatak and S. M. Weissman, *J. Virol.* **30**, 279 (1979).
304. J. A. Thomson, M. F. Radonovich and N. P. Salzmann, *J. Virol.* **31**, 437 (1979).
305. E. Paucha, A. Mellor, R. Harvey, A. E. Smith, R. M. Hewick and M. D. Waterfield, *PNAS* **75**, 2165 (1978).
306. N. Bouck, N. Beales, T. Shenk, P. Berg and G. D. Mayorca, *PNAS* **75**, 2473 (1978).
307. E. Paucha and A. E. Smith, *Cell* **15**, 1011 (1978).
308. E. Paucha, R. Harvey, R. Smith and A. E. Smith, *INSERM Colloq.* **69**, 189 (1977).
309. P. K. Ghosh, V. B. Reddy, J. Swinscoe, P. Lebowitz and S. M. Weissman, *JMB* **126**, 813 (1978).
310. Y. Aloni, R. Dhar, O. Laub, M. Horowitz and G. Khoury, *PNAS* **74**, 3686 (1977).
311. M. T. Hsu and J. Ford, *PNAS* **74**, 4982 (1977).
312. M. Bina-Stein, M. Thoren, N. Salzman and J. A. Thompson, *PNAS* **76**, 731 (1979).
313. G. Haegeman, H. V. Heuverswyn, D. Gheysen and W. Fiers, *J. Virol.* **31**, 484 (1979).
314. N. H. Chiu, M. F. Radonovich, M. M. Thoren and N. P. Salzman, *J. Virol.* **28**, 590 (1978).
315. H. Hamada, T. Igarashi and M. Muramatsu, *NARes* **8**, 587 (1980).
315a. M. Horowitz, O. Laub, S. Bratosin and Y. Aloni, *Nature* **275**, 558 (1978).
316. Y. Aloni, This volume.
317. I. Seif, G. Khoury and R. Dhar, *NARes* **6**, 3387 (1979).
318. N. H. Chiu, W. B. Bruszewski and N. P. Salzman, *NARes* **8**, 153 (1980).
319. G. Khoury, P. Gruss, R. Dhar, and C.-J. Lai, *Cell* **18**, 85 (1979).
320. M. Bina, R. J. Feldmann and R. G. Deeley, *PNAS* **77**, 1278 (1980).
321. M. R. Lerner, J. A. Boyle, S. M. Mount, S. L. Wolin and J. A. Steitz, *Nature* **283**, 220 (1980).
322. J. Rogers and R. Wall, *PNAS* **77**, 1877 (1980).

323. W. Jelinek and L. Leinwand, *Cell* **15**, 205 (1978).
324. B. W. Baer and R. D. Kornberg, *PNAS* **77**, 1890 (1980).
325. C.-J. Lai and G. Khoury, *PNAS* **76**, 71 (1979).
326. P. Gruss, C. J. Lai, R. Dhar and G. Khoury, *PNAS* **76**, 4317 (1979).
327. D. H. Hamer and P. Leder, *Cell* **18**, 1299 (1979).
328. L. T. Chow and T. R. Broker, *Cell* **15**, 497 (1978).
329. A. Kinniburgh and J. Ross, *Cell* **17**, 915 (1979).
330. C. Benoist, K. O'Hare, R. Breathnach and P. Chambon, *NARes* **8**, 127 (1980).
331. M. Levitt, *PNAS* **75**, 640 (1978).
332. D. Mathis, P. Oudet and P. Chambon, This Series, **24**, 1 (1980).
333. S. L. McKnight, M. Bustin and O. L. Miller, Jr., *CSHSQB* **42**, 741 (1978).
334. W. W. Franke, U. Scheer, M. Trendelenburg, H. Zentgraf and H. Spring, *CSHSQB* **42**, 755 (1978).
335. M. M. Lamb and B. Daneholt, *Cell* **17**, 835 (1979).
336. I. L. Goldknopf and H. Busch, in "The Cell Nucleus" (H. Busch, ed.), Vol. 6, pp. 149. Academic Press, New York, 1978.
337. S. Weisbrod and H. Weintraub, *PNAS* **76**, 630 (1979).
338. M.-J. Tsai, S. Y. Tsai and B. W. O'Malley, in "The Cell Nucleus" (H. Busch, ed.), Vol. 7, p. 163. Academic Press, New York, 1979.
339. M. Coca-Prados, G. Vitali and M. T. Hsu, *J. Virol.* **36**, 353 (1980).
340. R. G. Martin and V. P. Setlow, *Cell* **20**, 381 (1980).
341. O. Sundin and A. Varshavsky, *Cell* **21**, 103 (1980).
342. G. Haegeman and W. Fiers, *J. Virol.* **35**, 955 (1980).
343. D. Rio, A. Robbins, R. Myers and R. Tjian, *PNAS* **77**, 5706 (1980).
344. Y. Gluzman, J. F. Sambrook and R. J. Frisque, *PNAS* **77**, 3898 (1980).
345. C. Benoist and P. Chambon, *PNAS* **77**, 3865 (1980).

NOTE ADDED IN PROOF

Histones in mature virions and virion assembly intermediates are modified to a much greater extent than are those in SV40 and cellular chromatin. The modifications occur before the accumulation of capsid proteins around the SV40 chromatin (*339*), and may be necessary for efficient interaction between nucleosomes and capsid proteins for packaging of the SV40 minichromosome. Histone modification may be the mechanism by which the pool of minichromosomes destined for the synthesis of viral RNA or DNA can be distinguished from that for the production of progeny virus.

The uniqueness of the SV40 replication origin has been questioned (*340*). Over 98% of the replicative intermediates isolated from cells infected with wild-type virions at 33°, 37°, or 40°C and with *ts*A 209 at 33°C initiate replication near the *Bgl*I site (about 0.67 map unit), while 1% initiate replication about 2400 nucleotides away from the *Bgl*I site. Nonspecific initiation is a rare event in normal infection and is the major event when cells infected with *ts*A mutants are shifted to restrictive temperatures, the reasons for which are not clear.

An analysis of segregation intermediates shows that the terminal stages of SV40 DNA replication proceed via multiply intertwined catenated dimers, with the linkage states varying from 1 to 10 (*341*).

Similar to SV40 late mRNAs, as many as six capped structures are identified in early mRNAs from cells infected with SV40 wild type or *ts*A 209 mutant (*342*; Y. Groner, personal communication).

Cell-free transcription with cloned DNA templates and RNA polymerase II shows that a direct interaction between large T and its specific binding sites on SV40 DNA is the mechanism by which the SV40 early gene product autoregulates its transcription (343).

Some origin-defective mutants that have suffered deletions ranging from 4 to 241 nucleotides allow T antigen synthesis, and others do not, irrespective of the size or location of deletions. The 5' ends of the viral mRNAs are located about 25 ± 2 nucleotides downstream from the (A + T)-rich region (Hogness-Golberg box), indicating that this region serves as the promotor (234; P. K. Ghosh, personal communication). Deletions covering the putative promoter region for SV40 early mRNAs do not abolish large-T expression (345). This suggests that the (A + T)-rich region is only a part of the early promoter and its deletion does not completely eliminate promoter function.

Index

A

Alkaline sucrose gradient analysis, blockage of DNA synthesis at pyrimidine dimers and, 82–85
Amino acid(s), deprivation, protein synthesis and, 131
AUG, use in protein synthesis, 145–146

B

Bacteria
  mutagenesis by ultraviolet radiation, 100–101
    in derivatives of *lex* A and *rec* A mutants, 102–104
    other types, 105–106
    in repair-deficient mutants, 101–102
    SOS hypothesis, 101
    tandem base-pair changes and, 104–105
  repair-deficient mutants of, 58–60
Bacteriophage, repair of, 108–110
Base-pair, tandem changes, mutagenesis and, 104–105

C

Cell(s)
  differentiation, methylated bases in DNA and, 43–46
  mammalian
    mutagenesis by ultraviolet radiation, 100–108
    repair-deficient mutants of, 58–60
Cell cycle, ultraviolet-irradiation and, 77
Cell density, protein synthesis and, 131–132
Chromatin, eukaryotic, minichromosome as model for structure and function of, 228–232
Chromosomal replication, initiation, ultraviolet-irradiation and, 69–70
Chromosome
  distribution of methylated bases along
    chromosomal protein and, 39–40
    chromosome ultrastructure and, 40
    DNA sequences and, 38–39
    methylated and unmethylated domains, 40
  SV40 genome, artifacts of preparation, 195–199

D

Deoxyribonucleic acid
  control of replication in UV-irradiated mammalian cells, 75
  pairing of homologous strands, *rec* A gene and, 71
  repair
    rec A gene and, 70–71
    role of genetic recombination in, 92–95
  repair mechanisms, overview of, 56–58
  replication, methylation and, 47–48
  replication and mutagenesis, scope of review, 54–56
  segregation of histones along strands, 210–211
  SV40, initiation of transcription late after infection, 4–5
  SV40 genome
    initiation and termination of replication, 199–202
    mechanism of chain elongation, 202–203
Deoxyribonucleic acid polymerase, SV40 genome and, 205–206
Deoxyribonucleic acid synthesis
  effects of ultraviolet irradiation *in vitro*, 116
  methods of measurement of replication fidelity, 118–121
  replication of ultraviolet-damaged DNA, 117–118
  in ultraviolet-irradiated bacteria
    effects on initiation of chromosomal replication, 69–70
    evidence for continuous DNA synthesis past pyrimidine dimers *in vivo*, 67–69
    evidence for gaps opposite pyrimidine dimers, 61–63
    gaps in replicated DNA, 60–61
    inhibition of synthesis, 60

243

model for role of *rec* A and *lex* A genes in DNA repair, 70–74
recombinational mechanism for removal of gaps, 63–66
recovery of the ability to synthesize high-molecular-weight DNA, 66–67
in ultraviolet-irradiated mammalian cells
control of DNA replication, 75–77
does genetic recombination play a role in DNA repair or in mechanisms of tolerance?, 92–95
do gaps exist in replicated DNA opposite pyrimidine dimers?, 90–92
effects on the cell cycle, 77
evidence for and against inducible mechanism for replication of damaged DNA, 98–100
evidence for blockage of DNA synthesis at pyrimidine dimers, 82–89
inhibition of DNA elongation, 81–82
inhibition of DNA synthesis, 77–79
inhibition of replicon initiation, 79–81
mechanism for replication of damaged DNA, 96–98
replication past sites of DNA damage, 89–90

E

Endonuclease activity, (2'-5')-oligoadenylate-activated, 173
Energy, requirements for mRNA binding, 146

F

Fiber autoradiography, blockage of DNA synthesis at pyrimidine dimers and, 85–86

G

Gene(s), activity, methylated bases in DNA and, 43–46
Gene product(s), SV40 genome and, 203–207
Genetic recombination, role in DNA repair or tolerance of UV damage, 92–95
Globin
regulation of mRNA translation by hemin
historical background, 153–154
inhibitor formation, 158–161
phosphorylation of eIF-2a: role in phosphatase activity, 161–163
relationship to eIF-2a phosphorylation, 163–167
reticulocyte lysate, 154–158
reversal factors, 167
summary of mechanism of hemin control of translation, 167–169
Glucose, deprivation, protein synthesis and, 131

H

Hemin
regulation of globin mRNA translation by
historical background, 153–154
inhibitor formation, 158–161
phosphorylation of eIF-2a: role in phosphatase activity, 161–163
relationship to eIF-2a phosphorylation, 163–167
reticulocyte lysate, 154–158
reversal factors, 167
summary of mechanism of hemin control of translation, 167–169
Histones, assembly on SV40 genome, 207–208
segregation along DNA strands, 210–211
segregation in nucleosomes, 208–210
structure of nucleosomes, 208

I

Inhibitor formation, globin mRNA translation and, 158–161
Interferon
regulation of translation by
double-stranded RNA-dependent protein kinase(s), 170–171
historical background, 169–170

INDEX 245

(2′-5′)-oligoadenylate-activated endonuclease activity, 173
(2′-5′)-oligoadenylate and (2′-5′)-oligoadenylate synthase and, 171–172
(2′-5′)-phosphodiesterase activity, 173
relative importance of double-standard RNA-dependent enzyme activity, 173–174

L

lex A gene, role in DNA repair, 70–74
lex A mutants, mutagenesis in, 102–104

M

Methylases, specificity
  sequence, 35–38
  substrate, 34–35
Methylated bases
  distribution along chromosome
    methylated and unmethylated domains, 40
    with respect to chromosomal protein, 39–40
    with respect to chromosome ultrastructure, 40
    with respect to DNA sequences, 38–39
  possible functions of, 42–43
    cell differentiation and gene activity, 43–46
    interplay between DNA replication and methylation, 47–48
    mutation, recombination and repair, 48–49
    restriction and modification, 46–47
Methylation, modes *in vivo*, 40–41
  de novo, 41
  origins of, 42
  semiconservative, 41–42
Met-tRNA$_f$, binding to 43S$_N$ ribosomal subunits
  additional cofactors, 141–142
  properties of eIF-2, 137–139
  ternary complex binding, 141
  ternary complex formation, 139–141

Minichromosome
  as model for structure and function of eukaryotic chromatin, 228–232
  of SV40, 5–7
Minute virus of mice, late mRNAs, splicing of, 17–21
Moloney murine leukemia virus RNA, splicing, 21–22
  duplexes between cDNA and poly(A) containing RNA, 22–25
  duplexes between genomic RNA and cDNA, 22
Mutagenesis, by ultraviolet radiation
  in bacteria, 100–106
  in mammalian cells, 106–108
Mutation, methylated bases in DNA and, 48–49

N

Nucleosomes
  distribution, SV40 genome and, 192–193
  segregation of histones in, 208–210
  structure of, 208

O

(2′-5′)-Oligoadenylate synthase, (2′-5′)-oligoadenylate and, 171–172

P

Phosphatase activity, role of phosphorylation of eIF-2a in, 161–163
(2′-5′)-Phosphodiesterase activity, regulation of translation and, 173
Phytohemagglutinin, effects in T lymphocytes, 132
Polyoma, late mRNAs, splicing of, 16–17
Preinitiation complexes, mRNA-containing, evidence supporting *in vivo* existence, 147–148
Protein(s)
  binding mRNA, 142–145
  of SV40 genome, 189–191
Protein synthesis
  regulation
    importance of initiation in, 129–131
      amino-acid deprivation, 131
      effects of cell density and serum, 131–132

effects of phytohemagglutinin in T lymphocytes, 132
glucose deprivation, 131
regulation of initiation
of globin mRNA translation by hemin, 153–169
of translation by interferon, 169–174
relationship to eIF-2a phosphorylation, 163–167
sequence of events, 133–135
binding of mRNA to $43S_N$ preinitiation complex, 142–151
met-tRNA$_f$ binding to $43S_N$ ribosomal subunits, 137–142
production of native 43S ribosomal subunits, 135–137
ribosomal subunit joining and initiation factor release, 151–153
Pyrimidine dimers
continuous DNA synthesis past *in vivo*, 67–69
evidence for blockage of DNA synthesis at, 82–89
evidence for gaps opposite, 61–63

R

rec A gene
protein
regulation of synthesis of, 71–72
role in DNA repair, 70–74
role in UV-induced genetic exchanges, 72–74
rec A mutants, mutagenesis in, 102–104
Recombination, methylated bases in DNA and, 48–49
Repair, methylated bases in DNA and, 48–49
Replication complex, model for inhibition at damaged sites, 87–89
Replication fidelity, measurement of, 117–118
Replicon initiation, inhibition of, 79–81
Reticulocyte lysate, globin mRNA translation and, 154–158
Reversal factors, globin mRNA translation and, 167
Ribonucleic acid
double-stranded
protein kinase(s) and, 170–171
relative importance to dependent enzyme activity, 173–174
Ribonucleic acid
messenger
binding to $43S_N$ preinitiation complex, 142–151
existence as ribonucleoprotein complexes, 148–150
models for splicing, 25–27
processing of early, 224–225
processing of late, 225–226
recognition during initiation, 146–147
specificity factors, 150–151
spliced, techniques for analyzing, 15–16
Ribonucleic acid polymerase, transcriptional control of gene expression, SV40 genome and, 216–217
Ribosomal subunits
binding of met-tRNA$_f$ to, 137–142
joining, initiation factor release and, 151–153
native, production of, 135–137

S

Serum, protein synthesis and, 131–132
SOS hypothesis, bacterial mutagenesis and, 101
SV40
late mRNAs, splicing of, 7–9
minichromosome of, 5–7
SV40 DNA, initiation of transcription late after infection, 4–5
SV40 genome
assembly of histones on 207–208
segregation along DNA strands, 210–211
segregation in nucleosomes, 208–210
structure of nucleosomes, 208
gene product(s) and enzymes involved in DNA replication
DNA polymerases, 205–206
other enzymes and stimulatory factors, 206–207
T antigen, 203–205
posttranscriptional control, 223–224
mechanism of splicing, 226–228
processing of early mRNAs, 224–225
processing of late mRNAs, 225–226

INDEX

replication of DNA
  initiation and termination of, 199–202
  mechanism of chain elongation, 202–203
  structure
    artifacts of chromosome preparation, 195–199
    biophysical properties, 189
    heterogeneity, 194–195
    models of superstructure, 199
    protein composition, 189–191
    repeat length and distribution of nucleosomes, 192–193
  transcriptional control of gene expression
    promoter and terminator signals, 215–216
    RNA polymerase and other protein factors, 216–217
    role of T antigen, 221–223
    template, 218–221
  transcription of, 211–212
    early, 212–214
    late, 214–215

T

T antigen
  role in transcriptional control of gene expression of SV40 genome, 221–223
  SV40 genome and, 203–205
Template, transcriptional control of gene expression of SV40 genome and, 218–221
T lymphocytes, effects of phytohemagglutinin on, 132

U

Ultraviolet damage, mechanism of tolerance to, genetic recombination and, 92–95
Ultraviolet radiation, mutagenesis by
  in bacteria, 100–106
  in mammalian cells, 106–108

V

Viral mRNA(s)
  mapping of leader and body of by electron microscopy, 9–10
    analysis of DNA-RNA hybrids, 10–11
    analysis of R-loop structures, 11–14
    models for joining leader to coding sequences, 14–15
  splicing intermediates, 27–28
  splicing of, SV40 as model system, 2–4
Viral replication, studies of, 86–87
Virus, reactivation of ultraviolet-damaged
  mutagenesis associated with ultraviolet reactivation of animal viruses, 114–116
  mutagenesis of ultraviolet-irradiated bacteriophage, 110–112
  repair of animal viruses, 112–114
  repair of bacteriophage, 108–110

Contents of Previous Volumes

Volume 1
"Primer" in DNA Polymerase Reactions—F. J. Bollum
The Biosynthesis of Ribonucleic Acid in Animal Systems—R. M. S. Smellie
The Role of DNA in RNA Synthesis—Jerard Hurwitz and J. T. August
Polynucleotide Phosphorylase—M. Grunberg-Manago
Messenger Ribonucleic Acid—Fritz Lipmann
The Recent Excitement in the Coding Problem—F. H. C. Crick
Some Thoughts on the Double-Stranded Model of Deoxyribonucleic Acid—Aaron Bendich and Herbert S. Rosenkranz
Denaturation and Renaturation of Deoxyribonucleic Acid—J. Marmur, R. Rownd, and C. L. Schildkraut
Some Problems Concerning the Macromolecular Structure of Ribonucleic Acids—A. S. Spirin
The Structure of DNA as Determined by X-Ray Scattering Techniques—Vittorio Luzzati
Molecular Mechanisms of Radiation Effects—A. Wacker

Volume 2
Nucleic Acids and Information Transfer—Liebe F. Cavalieri and Barbara H. Rosenberg
Nuclear Ribonucleic Acid—Henry Harris
Plant Virus Nucleic Acids—Roy Markham
The Nucleases of *Escherichia coli*—I. R. Lehman
Specificity of Chemical Mutagenesis—David R. Krieg
Column Chromatography of Oligonucleotides and Polynucleotides—Matthys Staehelin
Mechanism of Action and Application of Azapyrimidines—J. Skoda
The Function of the Pyrimidine Base in the Ribonuclease Reaction—Herbert Witzel
Preparation, Fractionation, and Properties of sRNA—G. L. Brown

Volume 3
Isolation and Fractionation of Nucleic Acids—K. S. Kirby
Cellular Sites of RNA Synthesis—David M. Prescott
Ribonucleases in Taka-Diastase: Properties, Chemical Nature, and Applications—Fujio Egami, Kenji Takahashi, and Tsuneko Uchida
Chemical Effects of Ionizing Radiations on Nucleic Acids and Related Compounds—Joseph J. Weiss
The Regulation of RNA Synthesis in Bacteria—Frederick C. Neidhardt
Actinomycin and Nucleic Acid Function—E. Reich and I. H. Goldberg
De Novo Protein Synthesis *in Vitro*—B. Nisman and J. Pelmont
Free Nucleotides in Animal Tissues—P. Mandel

Volume 4
Fluorinated Pyrimidines—Charles Heidelberger
Genetic Recombination in Bacteriophage—E. Volkin
DNA Polymerases from Mammalian Cells—H. M. Keir
The Evolution of Base Sequences in Polynucleotides—B. J. McCarthy
Biosynthesis of Ribosomes in Bacterial Cells—Syozo Osawa
5-Hydroxymethylpyrimidines and Their Derivatives—T. L. V. Ulbricht
Amino Acid Esters of RNA, Nucleotides, and Related Compounds—H. G. Zachau and H. Feldmann
Uptake of DNA by Living Cells—L. Ledoux

Volume 5
Introduction to the Biochemistry of D-Arabinosyl Nucleosides—*Seymour S. Cohen*
Effects of Some Chemical Mutagens and Carcinogens on Nucleic Acids—*P. D. Lawley*
Nucleic Acids in Chloroplasts and Metabolic DNA—*Tatsuichi Iwamura*
Enzymatic Alteration of Macromolecular Structure—*P. R. Srinivasan and Ernest Borek*
Hormones and the Synthesis and Utilization of Ribonucleic Acids—*J. R. Tata*
Nucleoside Antibiotics—*Jack J. Fox, Kyoichi A. Watanabe, and Alexander Bloch*
Recombination of DNA Molecules—*Charles A. Thomas, Jr.*
   Appendix I. Recombination of a Pool of DNA Fragments with Complementary Single-Chain Ends—*G. S. Watson, W. K. Smith, and Charles A. Thomas, Jr.*
   Appendix II. Proof that Sequences of A, C, G, and T Can Be Assembled to Produce Chains of Ultimate Length, Avoiding Repetitions Everywhere—*A. S. Fraenkel and J. Gillis*
The Chemistry of Pseudouridine—*Robert Warner Chambers*
The Biochemistry of Pseudouridine—*Eugene Goldwasser and Robert L. Heinrikson*

Volume 6
Nucleic Acids and Mutability—*Stephen Zamenhof*
Specificity in the Structure of Transfer RNA—*Kin-ichiro Miura*
Synthetic Polynucleotides—*A. M. Michelson, J. Massoulié, and W. Guschlbauer*
The DNA of Chloroplasts, Mitochondria, and Centrioles—*S. Granick and Aharon Gibor*
Behavior, Neural Function, and RNA—*H. Hydén*
The Nucleolus and the Synthesis of Ribosomes—*Robert P. Perry*
The Nature and Biosynthesis of Nuclear Ribonucleic Acids—*G. P. Georgiev*
Replication of Phage RNA—*Charles Weissmann and Severo Ochoa*

Volume 7
Autoradiographic Studies on DNA Replication in Normal and Leukemic Human Chromosomes—*Felice Gavosto*
Proteins of the Cell Nucleus—*Lubomir S. Hnilica*
The Present Status of the Genetic Code—*Carl R. Woese*
The Search for the Messenger RNA of Hemoglobin—*H. Chantrenne, A. Burny, and G. Marbaix*
Ribonucleic Acids and Information Transfer in Animal Cells—*A. A. Hadjiolov*
Transfer of Genetic Information during Embryogenesis—*Martin Nemer*
Enzymatic Reduction of Ribonucleotides—*Agne Larsson and Peter Reichard*
The Mutagenic Action of Hydroxylamine—*J. H. Phillips and D. M. Brown*
Mammalian Nucleolytic Enzymes and Their Localization—*David Shugar and Halina Sierakowska*

Volume 8
Nucleic Acids—The First Hundred Years—*J. N. Davidson*
Nucleic Acids and Protamine in Salmon Testes—*Gordon H. Dixon and Michael Smith*
Experimental Approaches to the Determination of the Nucleotide Sequences of Large Oligonucleotides and Small Nucleic Acids—*Robert W. Holley*
Alterations of DNA Base Composition in Bacteria—*G. F. Gause*
Chemistry of Guanine and Its Biologically Significant Derivatives—*Robert Shapiro*
Bacteriophage φX174 and Related Viruses—*Robert L. Sinsheimer*
The Preparation and Characterization of Large Oligonucleotides—*George W. Rushizky and Herbert A. Sober*
Purine *N*-Oxides and Cancer—*George Bosworth Brown*
The Photochemistry, Photobiology, and Repair of Polynucleotides—*R. B. Setlow*
What Really is DNA? Remarks on the Changing Aspects of a Scientific Concept—*Erwin Chargaff*
Recent Nucleic Acid Research in China—*Tien-Hsi Cheng and Roy H. Doi*

CONTENTS OF PREVIOUS VOLUMES

Volume 9
The Role of Conformation in Chemical Mutagenesis—*B. Singer and H. Fraenkel-Conrat*
Polarographic Techniques in Nucleic Acid Research—*E. Paleček*
RNA Polymerase and the Control of RNA Synthesis—*John P. Richardson*
Radiation-Induced Alterations in the Structure of Deoxyribonucleic Acid and Their Biological Consequences—*D. T. Kanazir*
Optical Rotatory Dispersion and Circular Dichroism of Nucleic Acids—*Jen Tsi Yang and Tatsuya Samejima*
The Specificity of Molecular Hybridization in Relation to Studies on Higher Organisms—*P. M. B. Walker*
Quantum-Mechanical Investigations of the Electronic Structure of Nucleic Acids and Their Constituents—*Bernard Pullman and Alberte Pullman*
The Chemical Modification of Nucleic Acids—*N. K. Kochetkov and E. I. Budowsky*

Volume 10
Induced Activation of Amino Acid Activating Enzymes by Amino Acids and tRNA—*Alan H. Mehler*
Transfer RNA and Cell Differentiation—*Noboru Sueoka and Tamiko Kano-Sueoka*
N^6-(Δ^2-Isopentenyl)adenosine: Chemical Reactions, Biosynthesis, Metabolism, and Significance to the Structure and Function of tRNA—*Ross H. Hall*
Nucleotide Biosynthesis from Preformed Purines in Mammalian Cells: Regulatory Mechanisms and Biological Significance—*A. W. Murray, Daphne C. Elliott, and M. R. Atkinson*
Ribosome Specificity of Protein Synthesis *in Vitro*—*Orio Ciferri and Bruno Parisi*
Synthetic Nucleotide-peptides—*Zoe A. Shabarova*
The Crystal Structures of Purines, Pyrimidines and Their Intermolecular Complexes—*Donald Voet and Alexander Rich*

Volume 11
The Induction of Interferon by Natural and Synthetic Polynucleotides—*Clarence Colby, Jr.*
Ribonucleic Acid Maturation in Animal Cells—*R. H. Burdon*
Liporibonucleoprotein as an Integral Part of Animal Cell Membranes—*V. S. Shapot and S. Ya. Davidova*
Uptake of Nonviral Nucleic Acids by Mammalian Cells—*Pushpa M. Bhargava and G. Shanmugam*
The Relaxed Control Phenomenon—*Ann M. Ryan and Ernest Borek*
Molecular Aspects of Genetic Recombination—*Cedric I. Davern*
Principles and Practices of Nucleic Acid Hybridization—*David E. Kennell*
Recent Studies Concerning the Coding Mechanism—*Thomas H. Jukes and Lila Gatlin*
The Ribosomal RNA Cistrons—*M. L. Birnstiel, M. Chipchase, and J. Speirs*
Three-Dimensional Structure of tRNA—*Friedrich Cramer*
Current Thoughts on the Replication of DNA—*Andrew Becker and Jerard Hurwitz*
Reaction of Aminoacyl-tRNA Synthetases with Heterologous tRNA's—*K. Bruce Jacobson*
On the Recognition of tRNA by Its Aminoacyl-tRNA Ligase—*Robert W. Chambers*

Volume 12
Ultraviolet Photochemistry as a Probe of Polyribonucleotide Conformation—*A. J. Lomant and Jacques R. Fresco*
Some Recent Developments in DNA Enzymology—*Mehran Goulian*
Minor Components in Transfer RNA: Their Characterization, Location, and Function—*Susumu Nishimura*
The Mechanism of Aminoacylation of Transfer RNA—*Robert B. Loftfield*
Regulation of RNA Synthesis—*Ekkehard K. F. Bautz*
The Poly(dA-dT) of Crab—*M. Laskowski, Sr.*

The Chemical Synthesis and the Biochemical Properties of Peptidyl-tRNA—*Yehuda Lapidot and Nathan de Groot*

Volume 13
Reactions of Nucleic Acids and Nucleoproteins with Formaldehyde—*M. Ya. Feldman*
Synthesis and Functions of the -C-C-A Terminus of Transfer RNA—*Murray P. Deutscher*
Mammalian RNA Polymerases—*Samson T. Jacob*
Poly(adenosine diphosphate ribose)—*Takashi Sugimura*
The Stereochemistry of Actinomycin Binding to DNA and Its Implications in Molecular Biology—*Henry M. Sobell*
Resistance Factors and Their Ecological Importance to Bacteria and to Man—*M. H. Richmond*
Lysogenic Induction—*Ernest Borek and Ann Ryan*
Recognition in Nucleic Acids and the Anticodon Families—*Jacques Ninio*
Translation and Transcription of the Tryptophan Operon—*Fumio Imamoto*
Lymphoid Cell RNA's and Immunity—*A. Arthur Gottlieb*

Volume 14
DNA Modification and Restriction—*Werner Arber*
Mechanism of Bacterial Transformation and Transfection—*Nihal K. Notani and Jane K. Setlow*
DNA Polymerases II and III of *Escherichia coli*—*Malcolm L. Gefter*
The Primary Structure of DNA—*Kenneth Murray and Robert W. Old*
RNA-Directed DNA Polymerase—Properties and Functions in Oncogenic RNA Viruses and Cells—*Maurice Green and Gary F. Gerard*

Volume 15
Information Transfer in Cells Infected by RNA Tumor Viruses and Extension to Human Neoplasia—*D. Gillespie, W. C. Saxinger, and R. C. Gallo*
Mammalian DNA Polymerases—*F. J. Bollum*
Eukaryotic RNA Polymerases and the Factors That Control Them—*B. B. Biswas, A. Ganguly, and D. Das*
Structural and Energetic Consequences of Noncomplementary Base Oppositions in Nucleic Acid Helices—*A. J. Lomant and Jacques R. Fresco*
The Chemical Effects of Nucleic Acid Alkylation and Their Relation to Mutagenesis and Carcinogenesis—*B. Singer*
Effects of the Antibiotics Netropsin and Distamycin A on the Structure and Function of Nucleic Acids—*Christoph Zimmer*

Volume 16
Initiation of Enzymic Synthesis of Deoxyribonucleic Acid by Ribonucleic Acid Primers—*Erwin Chargaff*
Transcription and Processing of Transfer RNA Precursors—*John D. Smith*
Bisulfite Modification of Nucleic Acids and Their Constituents—*Hikoya Hayatsu*
The Mechanism of the Mutagenic Action of Hydroxylamines—*E. I. Budowsky*
Diethyl Pyrocarbonate in Nucleic Acid Research—*L. Ehrenberg, I. Fedorcsák, and F. Solymosy*

Volume 17
The Enzymic Mechanism of Guanosine 5′, 3′-Polyphosphate Synthesis—*Fritz Lipmann and Jose Sy*
Effects of Polyamines on the Structure and Reactivity of tRNA—*Ted T. Sakai and Seymour S. Cohen*

CONTENTS OF PREVIOUS VOLUMES 253

Information Transfer and Sperm Uptake by Mammalian Somatic Cells—*Aaron Bendich, Ellen Borenfreund, Steven S. Witkins, Delia Beju, and Paul J. Higgins*
Studies on the Ribosome and Its Components—*Pnina Spitnik-Elson and David Elson*
Classical and Postclassical Modes of Regulation of the Synthesis of Degradative Bacterial Enzymes—*Boris Magasanik*
Characteristics and Significance of the Polyadenylate Sequence in Mammalian Messenger RNA—*George Brawerman*
Polyadenylate Polymerases—*Mary Edmonds and Mary Ann Winters*
Three-Dimensional Structure of Transfer RNA—*Sung-Hou Kim*
Insights into Protein Biosynthesis and Ribosome Function through Inhibitors—*Sidney Pestka*
Interaction with Nucleic Acids of Carcinogenic and Mutagenic N-Nitroso Compounds—*W. Lijinsky*
Biochemistry and Physiology of Bacterial Ribonuclease—*Alok K. Datta and Salil K. Niyogi*

Volume 18

The Ribosome of *Escherichia coli*—*R. Brimacombe, K. H. Nierhaus, R. A. Garrett and H. G. Wittmann*
Structure and Function of 5 S and 5.8 S RNA—*Volker A. Erdmann*
High-Resolution Nuclear Magnetic Resonance Investigations of the Structure of tRNA in Solution—*David R. Kearns*
Premelting Changes in DNA Conformation—*E. Paleček*
Quantum-Mechanical Studies on the Conformation of Nucleic Acids and Their Constituents—*Bernard Pullman and Anil Saran*

Volume 19 (Symposium on mRNA: The Relation of Structure to Function)

I. The 5'-Terminal Sequence ("Cap") of mRNAs
Caps in Eukaryotic mRNAs: Mechanism of Formation of Reovirus mRNA 5'-Terminal m⁷GpppGm-C—*Y. Furuichi, S. Muthukrishnan, J. Tomasz and A. J. Shatkin*
Nucleotide Methylation Patterns in Eukaryotic mRNA—*Fritz M. Rottman, Ronald C. Desrosiers and Karen Friderici*
Structural and Functional Studies on the "5'-Cap": A Survey Method of mRNA—*Harris Busch, Friedrich Hirsch, Kaushal Kumar Gupta, Manchanahalli Rao, William Spohn and Benjamin C. Wu*
Modification of the 5'-Terminals of mRNAs by Viral and Cellular Enzymes—*Bernard Moss, Scott A. Martin, Marcia J. Ensinger, Robert F. Boone and Cha-Mer Wei*
Blocked and Unblocked 5' Termini in Vesicular Stomatitis Virus Product RNA *in Vitro*: Their Possible Role in mRNA Biosynthesis—*Richard J. Colonno, Gordon Abraham and Amiya K. Banerjee*
The Genome of Poliovirus Is an Exceptional Eukaryotic mRNA—*Yuan Fon Lee, Akio Nomoto and Eckard Wimmer*
II. Sequences and Conformations of mRNAs
Transcribed Oligonucleotide Sequences in Hela Cell hnRNA and mRNA—*Mary Edmonds, Hiroshi Nakazato, E. L. Korwek and S. Venkatesan*
Polyadenylylation of Stored mRNA in Cotton Seed Germination—*Barry Harris and Leon Dure III*
mRNAs Containing and Lacking Poly(A) Function as Separate and Distinct Classes during Embryonic Development—*Martin Nemer and Saul Surrey*
Sequence Analysis of Eukaryotic mRNA—*N. J. Proudfoot, C. C. Cheng and G. G. Brownlee*
The Structure and Function of Protamine mRNA from Developing Trout Testis—*P. L. Davies, G. H. Dixon, L. N. Ferrier, L. Gedamu and K. Iatrou*
The Primary Structure of Regions of SV40 DNA Encoding the Ends of mRNA—*Kiranur N.*

Subramanian, Prabhat K. Ghoshi, Ravi Dhar, Bayar Thimmappaya, Sayeeda B. Zain, Julian Pan and Sherman M. Weissman

Nucleotide Sequence Analysis of Coding and Noncoding Regions of Human β-Globin mRNA—*Charles A. Marotta, Bernard G. Forget, Michael Cohen/Solal and Sherman M. Weissman*

Determination of Globin mRNA Sequences and Their Insertion into Bacterial Plasmids—*Winston Salser, Jeff Browne, Pat Clarke, Howard Heindell, Russell Higuchi, Gary Paddock, John Roberts, Gary Studnicka and Paul Zakar*

The Chromosomal Arrangement of Coding Sequences in a Family of Repeated Genes—*G. M. Rubin, D. J. Finnegan and D. S. Hogness*

Mutation Rates in Globin Genes: The Genetic Load and Haldane's Dilemma—*Winston Salser and Judith Strommer Isaacson*

Heterogeneity of the 3' Portion of Sequences Related to Immunoglobulin κ-Chain mRNA—*Ursula Storb*

Structural Studies on Intact and Deadenylylated Rabbit Globin mRNA—*John N. Vournakis, Marcia S. Flashner, MaryAnn Katopes, Gary A. Kitos, Nikos C. Vamvakopoulos, Matthew S. Sell and Regina M. Wurst*

Molecular Weight Distribution of RNA Fractionated on Aqueous and 70% Formamide Sucrose Gradients—*Helga Boedtker and Hans Lehrach*

III. Processing of mRNAs

Bacteriophages T7 and T3 as Model Systems for RNA Synthesis and Processing—*J. J. Dunn, C. W. Anderson, J. F. Atkins, D. C. Bartelt and W. C. Crockett*

The Relationship between hnRNA and mRNA—*Robert P. Perry, Enzo Bard, B. David Hames, Dawn E. Kelley and Ueli Schibler*

A Comparison of Nuclear and Cytoplasmic Viral RNAs Synthesized Early in Productive Infection with Adenovirus 2—*Heschel J. Raskas and Elizabeth A. Craig*

Biogenesis of Silk Fibroin mRNA: An Example of Very Rapid Processing?—*Paul M. Lizardi*

Visualization of the Silk Fibroin Transcription Unit and Nascent Silk Fibroin Molecules on Polyribosomes of *Bombyx mori*—*Steven L. McKnight, Nelda L. Sullivan and Oscar L. Miller, Jr.*

Production and Fate of Balbiani Ring Products—*B. Daneholt, S. T. Case, J. Hyde, L. Nelson and L. Wieslander*

Distribution of hnRNA and mRNA Sequences in Nuclear Ribonucleoprotein Complexes—*Alan J. Kinniburgh, Peter B. Billings, Thomas J. Quinlan and Terence E. Martin*

IV. Chromatin Structure and Template Activity

The Structure of Specific Genes in Chromatin—*Richard Axel*

The Structure of DNA in Native Chromatin as Determined by Ethidium Bromide Binding—*J. Paoletti, B. B. Magee and P. T. Magee*

Cellular Skeletons and RNA Messages—*Ronald Herman, Gary Zieve, Jeffrey Williams, Robert Lenk and Sheldon Penman*

The Mechanism of Steroid-Hormone Regulation of Transcription of Specific Eukaryotic Genes—*Bert W. O'Malley and Anthony R. Means*

Nonhistone Chromosomal Proteins and Histone Gene Transcription—*Gary Stein, Janet Stein, Lewis Kleinsmith, William Park, Robert Jansing and Judith Thomson*

Selective Transcription of DNA Mediated by Nonhistone Proteins—*Tung Y. Wang, Nina C. Kostraba and Ruth S. Newman*

V. Control of Translation

Structure and Function of the RNAs of Brome Mosaic Virus—*Paul Kaesberg*

Effect of 5'-Terminal Structures on the Binding of Ribopolymers to Eukaryotic Ribosomes—*S. Muthukrishnan, Y. Furuichi, G. W. Both and A. J. Shatkin*

CONTENTS OF PREVIOUS VOLUMES 255

Translational Control in Embryonic Muscle—*Stuart M. Heywood and Doris S. Kennedy*
Protein and mRNA Synthesis in Cultured Muscle Cells—*R. G. Whalen, M. E. Buckingham and F. Gros*
VI. Summary: mRNA Structure and Function—*James E. Darnell*

Volume 20

Correlation of Biological Activities with Structural Features of Transfer RNA—*B. F. C. Clark*
Bleomycin, an Antibiotic That Removes Thymine from Double-Stranded DNA—*Werner E. G. Müller and Rudolf K. Zahn*
Mammalian Nucleolytic Enzymes—*Halina Sierakowska and David Shugar*
Transfer RNA in RNA Tumor Viruses—*Larry C. Waters and Beth C. Mullin*
Integration versus Degradation of Exogenous DNA in Plants: An Open Question—*Paul F. Lurquin*
Initiation Mechanisms of Protein Synthesis—*Marianne Grunberg-Manago and François Gros*

Volume 21

Informosomes and Their Protein Components: The Present State of Knowledge—*A. A. Preobrazhensky and A. S. Spirin*
Energetics of the Ribosome—*A. S. Spirin*
Mechanisms in Polypeptide Chain Elongation on Ribosomes—*Engin Bermek*
Synthetic Oligodeoxynucleotides for Analysis of DNA Structure and Function—*Ray Wu, Chander P. Bahl and Saran A. Narang*
The Transfer RNAs of Eukaryotic Organelles—*W. Edgar Barnett, S. D. Schwartzbach, and L. I. Hecker*
Regulation of the Biosynthesis of Aminoacid:tRNA Ligases and of tRNA—*Susan D. Morgan and Dieter Söll*

Volume 22

The -C-C-A End of tRNA and Its Role in Protein Biosynthesis—*Mathias Sprinzl and Friedrich Cramer*
The Mechanism of Action of Antitumor Platinum Compounds—*J. J. Roberts and A. J. Thomson*
DNA Glycosylases, Endonucleases for Apurinic/Apyrimidinic Sites, and Base Excision-Repair—*Thomas Lindahl*
Naturally Occurring Nucleoside and Nucleotide Antibiotics—*Robert J. Suhadolnik*
Genetically Controlled Variation in the Shapes of Enzymes—*George Johnson*
Transcription Units for mRNA Production in Eukaryotic Cells and Their DNA Viruses—*James E. Darnell, Jr.*

Volume 23

The Peptidyltransferase Center of Ribosomes—*Alexander A. Krayevsky and Marina K. Kukhanova*
Patterns of Nucleic Acid Synthesis in *Physarum polycephalum*—*Geoffrey Turnock*
Biochemical Effects of the Modification of Nucleic Acids by Certain Polycyclic Aromatic Carcinogens—*Dezider Grunberger and I. Bernard Weinstein*
Participation of Modified Nucleosides in Translation and Transcription—*B. Singer and M. Kröger*
The Accuracy of Translation—*Michael Yarus*
Structure, Function, and Evolution of Transfer RNAs (with Appendix Giving Complete Sequences of 178 tRNAs)—*Ram P. Singhal and Pamela A. M. Fallis*

Volume 24

Structure of Transcribing Chromatin—*Diane Mathis, Pierre Oudet, and Pierre Chambon*
Ligand-Induced Conformational Changes in Ribonucleic Acids—*Hans Günter Gassen*
Replicative DNA Polymerases and Mechanisms at a Replication Fork—*Robert K. Fujimura and Shishir K. Das*
Antibodies Specific for Modified Nucleosides: An Immunochemical Approach for the Isolation and Characterization of Nucleic Acids—*Theodore W. Munns and M. Kathryn Liszewski*
DNA Structure and Gene Replication—*R. D. Wells, T. C. Goodman, W. Hillen, G. T. Horn, R. D. Klein, J. E. Larson, U. R. Müller, S. K. Neuendorf, N. Panayotatos, and S. M. Stirdivant*